Unmanned Air Vehicles

An Illustrated Study of UAVs

Bill Holder

Schiffer Military/Aviation History
Atglen, PA

Acknolwedgments

1. Kaye LeBeFebure, Scaled Composites
2. Melissa Doboski, Israel Aircraft Industries, Inc.
3. Alan Brown, NASA Dryden Flight Research Center
4. Craig Ballard, Pioneer UAV Inc.
5. David Lanman, AFRL/VA, USAF
6. Gary Geiger, Alliant Tech Systems
7. Cynthia Curiel, Northrop Grumman Corp.
8. Bruce Hess, ASC History Office, USAF
9. Ed Steadham & James Cycon, Sikorsky Aircraft
10. Jay Willmott, BAI Aerosystems

Book Design by Ian Robertson.

Library of Congress Control Number: 2001094439

Printed in China.
ISBN: 0-7643-1500-5

We are interested in hearing from authors with book ideas on related topics.

Published by Schiffer Publishing Ltd.
4880 Lower Valley Road
Atglen, PA 19310
Phone: (610) 593-1777
FAX: (610) 593-2002
E-mail: Schifferbk@aol.com.
Visit our web site at: www.schifferbooks.com
Please write for a free catalog.
This book may be purchased from the publisher.
Please include $3.95 postage.
Try your bookstore first.

In Europe, Schiffer books are distributed by:
Bushwood Books
6 Marksbury Avenue
Kew Gardens
Surrey TW9 4JF
England
Phone: 44 (0) 20 8392-8585
FAX: 44 (0) 20 8392-9876
E-mail: Bushwd@aol.com.
Free postage in the UK. Europe: air mail at cost.
Try your bookstore first.

Contents

Foreword

Unmanned Air Vehicles (UAVs) have been around for many years, starting with the Kettering Bug in 1918. In the past decade or so, UAVs have truly demonstrated their military utility, particularly in Desert Storm and the conflicts in the former Yugoslavia. Now, with UAVs gaining support from military leaders and showing great promise for commercial application, we are not only entering a new century, we are entering a new era of powered flight.

Bill's book does a great job of highlighting many of the major UAV programs, past and present, domestic and foreign. Bill also touches upon some of the development UAVs of the future. This book is not only a UAV reference document, but it also provides interesting history and background. A must for every aviation enthusiast's bookshelf!

David M. Lanman
President, Wright-Kettering Chapter
Association for Unmanned Vehicle
Systems International (AUVSI)

Introduction

UAV. It is the acronym of the new millenium, and could prove to be the greatest revolution in warfare in many, many decades. The symbol stands for Unmanned (or Uninhabited) Air (or Aerial) Vehicle.

The concept for the UAV in a military application is simple: design an UNMANNED vehicle to do that normally done by a MANNED aircraft.

Consider the implications of such a change. First, there is the matter of placing a pilot—or even worse, a crew—in harm's way. Also, should the UAV be lost, it is not necessary to mount an extensive and very dangerous rescue attempt, again putting additional friendlies in danger.

Certain societies do not rate the pilot's life worth such an effort, but the U.S. and its allies certainly do. You could, therefore, say that in certain situations, a UAV would be considered an expendable vehicle.

Another implication of the manned versus unmanned discussion is the fact that when the vehicle is inhabited, extensive life support equipment is required, equipment which consumes considerable volume and adds considerable weight. Therefore, to accomplish the same mission, all other factors being equal, it would appear that a much smaller, and probably less detectable, unmanned vehicle could be used. And in this era of decreasing military budgets, the economic advantage these vehicles provide is possibly their greatest advantage.

Surprisingly, just about every type of propulsion system has been considered to propell UAVs. They have used both single and multiple engines, along with piston, rotary, turbojet, pulsejet, or rocket engines. UAVs are used with and without boosters. Sizewise, the UAV family is just about as varied. The unmanned vehicles vary in size from systems that can be held in the palm of the hand, to vehicles that approach the size of full-size manned aircraft. They also have just about every appearance imaginable, from the look of a model airplane all the way to the appearance of an advanced stealth fighter or bomber.

The General Atomics Gnat 750 UAV is used by a number of domestic and foreign customers for a multitude of missions. (General Atomics Photo)

Right: This advanced UAV concept shows an aircraft-appearing vehicle with wingtip engines and a twin tail with the tails canting inward. The engine intake is located where the cockpit would normally be. (U.S. Navy Drawing)

Note the unique twin-wing configuration of the Lears II UAV. UAVs come in all sizes and shapes. (Bosch Photo)

Some UAVs began life as a manned fighter, such as this QF-4, which today are used as target drones. (McDonnell-Douglas Photo)

With the unmanned implications of the UAV, it becomes obvious that control for the vehicle must come from an outside source, either from the ground (or water) or from the air. The UAV can use many different types of guidance, such as pre-programmed, command, radar, radio, active seeker, etc, or a combination of a number of them.

UAVs performing missions in a military threat environment have been appropriately named UCAVs, for Unmanned Combat Air Vehicles. The UAV is more of a general military tool. Therefore, it is not surprising that within the U.S. military establishment, along with the military of other countries, that multiple services use the same UAV system.

Even though the UAV concept seems somewhat revolutionary in nature, it really isn't new, with a number of deviations of the concept being attempted for a number of decades.

One of the most common and well used versions of the UAV concept is in the use of unmanned drones as maneuverable targets for fighter aircraft guns and air-to-air missiles. Many times, older retired fighters have been used in this function, with fighter aircraft like the F-80, F-100, F-102, and more recently, the F-4 being used in that capacity. When used in this application, the target planes are given a Q designation, ie QF-4. These drones could, in a broad sense, be considered UAVs.

Of course, much of the technology required to control and maneuver these "Q" planes applies directly to the development of UAVs.

Not all drones were once manned vehicles, but instead are unmanned vehicles that were designed that way from scratch.

Through the years, the Air Force and Navy have used many different drones to hone the offensive skills of their fighter pilots.

The Ryan Firebee has been used for many years in many different versions in this capacity. The Firebee is jet-powered and parachute-recoverable. Initial versions of the Firebee were first used in the early 1950s.

The Ryan Firebee drone has been a mainstay with the U.S. military since the 1960s. (Ryan Photo)

If this vehicle looks something like a manned B-2 bomber, that would be true, since this Northrop configuration was a candidate for the Unmanned Combat Air Vehicle (UCAV) competition. (Northrop Photo)

UAV techniques are even being undertaken with manned aircraft. Here, a Navy F/A-18 makes a hands-off automatic landing, just as UAVs do. (Raytheon Photo)

The twin-tail Shadow 200 has been selected by the U.S. military to perform during the early decades of the 21st century. (AAI Photo)

Some UAVs look almost exactly like their manned counter-parts. This Northrop helicopter-appearing UAV could find its way into operational service in the future. (Northrop Photo)

The advanced supersonic Firebee II has a Mach 1.78 speed capability and a maximum altitude capability of over 60,000 feet. It should be noted that there have been a number of other drones built through the years by a number of manufacturers.

Moving to pure UAV developments, the numbers of different types have been staggering through the decades of the 1980s and 1990s, and into the new millennium. Of course, a majority of them did not reach beyond the research or test phase, but they certainly served to increase the body of technology for development of these unique vehicles.

It is extremely hard to classify the development of UAVs into distinct groups, but probably the best way is to consider them by the various U.S. armed services, other government agencies, and the multitude of foreign developments.

In the early years of the 21st century, the USAF was concentrating on development of the bulbous Global Hawk, a jet-powered UAV which will play greatly in Air Force surveil-

The NASA X-34 Technology Testbed Demonstrator is a one-of-a-kind UAV designed strictly for research purposes. The system investigated technologies for future low-cost reuseable launch vehicles. (NASA Photo)

The Freewing research UAV can actually tilt its body in flight. (Freewing Photo)

The Model 324 UAV is a U.S.-produced system that is used by the Egyptian Air Force. (Teledyne Ryan Photo)

The Mk 105 Flash UAV is a small pusher-prop powered vehicle that concentrates on commercial-type missions. (IAI Photo)

lance missions. The low-wing, fat-nosed Predator will also serve in important reconnaissance missions. One of the wildest-looking UAVs was the so-called DarkStar, a stealthy flat-body UAV that met the economic axe in the late 1990s.

It was mentioned about unmanned fighters being used for drone purposes. Well, the same concept could possibly evolve using the fighters in a future offensive role as an Unmanned Combat Air Vehicle. Then, there have also been studies of UCAVs that look like miniature versions of existing aircraft, such as one that looks almost exactly like the USAF B-2 bomber.

Along that same line of thinking, there has been some consideration for the next USAF bomber to be of the unmanned persuasion. In the fast-moving UAV arena, it is truly difficult to speculate what will happen in the future.

The U.S. Navy plans considerable operational use of UAVs in the future, with consideration of mostly shipboard use. Research has been conducted on the use of vertical take-off UAVs that could be launched from the confined decks of ships, therefore not requiring the use of any runway. The vehicles could provide reconnaissance capabilities now performed by manned aircraft. The service selected such a ve-

In the broadest sense, even turbojet-powered ballistic missiles like the 1960s-vintage Matador could fall into that UAV category. (Bill Holder Photo)

hicle, the helicopter-like Northrop Grumman FireScout, to fulfill this mission.

The Navy, like the Air Force, is also examined the possibility of modifying existing fighters into UCAV vehicles, along with the derivation of stealthy UCAV fighter configurations. Both of these UAV vehicle types could be launched from carrier decks.

With the aging of the EA-6B electronic warfare aircraft in the early 2000s, the Navy was also looking to the UAV to perform some of the E-6 airborne electronic attack and jamming missions.

The use of UAV guidance technologies was also applied to a manned aircraft. Navy testing has allowed an F/A-18 fighter to land on aircraft carriers automatically, possibly eliminating this most difficult maneuver in the future.

The USMC also has examinined the possibilties of using an unmanned helicopter for the delivery of supplies from offshore Navy ships to small, widely-dispersed Marine units ashore.

The U.S. Army, in 2000, fulfilled its reconnaissance requirements with the selection of the Shadow 200 UAV in the Tactical Unmanned Air Vehicle (TUAV) competition. In addition, the Army is also looking for UAV capabilities for the jamming of electronic signals.

The UAV has also been considered as major player in coordination of a total battle scenerio. A UAV could deploy in environments where data is needed quickly, manned aircraft are unavailable, or excessive risk or other conditions call for using an unmanned vehicle.

Besides the military services, the other big player in the development of military UAV vehicles is the DARPA, the U.S. Defense Advanced Research Projects Agency. DARPA is involved with a number of miniature UAVs that will have application with all services. Some of the amazing tiny vehicles can actually be held in the palm of a hand.

But the point must be made at this point that not all UAVs have military implications. The UAV also has numerous commercial applications, where a number of have been developed, and continue to be pursued, by the National Aeronatics and Space Administration (NASA) and other organizations.

NASA has also been researching the possibility of using UAVs for future communications networks. Japan has investigated the use of a helicopter-type UAV for crop spraying. Other UAV research activities in Europe have examined the use of UAVs for fish monitoring. The U.S. has also looked into the use of UAVs for U.S. border monitoring. In Japan, efforts have been directed toward the development of an agriculture monitoring UAV capability.

Other research has looked into the development of advanced engines built specifically for UAV applications. Much of the research has concerned the use of new fuels. There were also design efforts into the development of a parachute-based UAV, a vehicle which had a cruise speed of only 30 miles per hour.

There is also a considerable foreign UAV development effort. The popularity of the UAV concept has caused an amazing worldwide interest. Many countries have native designs, have formed consortiums with other countries, or have bought UAVs from other countries. This is definitely more than a passing fad.

Many other implications concerning the use of UAVs have been considered, including ways to actually refuel the unmanned vehicles at altitude, thus increasing time on station and range.

There has also been research work accomplished for automatic landing systems which could result in increased airframe lifetime.

There is a hazy line where UAVs actually start and cease to exist. For example, is an aerodynamic offensive cruise missile a UAV? It is a difficult question to answer, but in the broad sence, it probably could be included in the UAV family!

That similarity also brings about possible legal problems when a UAV is equipped with offensive weapons. Does its

Even the early surface-to-air Bomarc could roughly fall into the UAV arena. The lines are hazy, to say the least. (Bill Holder Photo)

use fall under the treaties that also govern cruise missiles? Only time will tell.

Another interesting concept investigated involved the use of a UAV to direct the flight path of a guided bomb instead of risking the use of a manned aircraft to do the job.

Guess it's not surprising to learn that UAVs are also considered as a valuable reconnaissance tool in urban war observation. With helicopters being highly vulnerable to ground fire, the UAV's small size makes it a very difficult target to engage. There have even been considerations for integrating UAVs as an aid to commercial traffic control.

It is hard to place a number on the actual count of UAV developments. Many, of course, went no further than an engineer's computer screen. Others reached the mock-up stage, while still others rearched a prototype flight vehicle, with some of those being flight tested. Then, there are those that were finally selected for operational service.

A little-known fact is that a large number of UAVs have already seen wartime service in a number of different conflicts, with systems from a number of different countries. And it would appear that this trend is just beginning.

Perhaps the ultimate use of a UAV was proposed by NASA for the 2003 time period. The space agency's 2000 budget included funds to develop a "Mars Unmanned Aircraft" which would fly over the red planet.

And the list goes on and on. Whether the UAV will end in overwhelming success, or possibly as a disappointment, remains to be seen. Only time will tell.

UAVs at War

	UAV System	Country	Year	Location of Use
1.	Seamos	Germany	1999	Kosovo
2.	Hunter	USA	1999	Kosovo
3.	QH-50	USA	1966	Gulf of Tonkin, Vietnam
4.	AGM-34	USA	1967-71	North Vietnam
5.	Firebee	Israel	1980s	Syria
6.	Firebee	Israel	1970s	Egypt
7.	Searcher	Israel	1996	Hizbollatt
8.	Pioneer	USA	1991	Desert Storm
9.	Searcher	India	1990s	SRI Lanka
10.	Kentron	S. Africa	1986	Angola
11.	Seeker	S. Africa	1994	Election monitoring
12.	La-17	Russia	1990s	—
13.	Mart Mark 1	France	1991	Desert Storm
14.	Pioneer	USA	1990s	Bosnia
15.	Gnat 750	USA	1990s	Bosnia
16.	Predator	USA	1995	Bosnia
17.	CL-289	France	1990s	Bosnia
18.	FOX AT1	France	1990s	Bosnia
19.	Pioneer	USA	1990s	Bosnia
20.	CL-289	Germany	1990s	Yugoslavia
21.	DR-3	Iraq	1980-88	Iran
22.	Pointer	USA	1991	Desert Storm
23,	Chukar 3	USA	1991	Desert Storm
24.	Pheonix	Britain	1990s	Kosovo

In 2003, there might even be a UAV flying over the Mars surface. (NASA Sketch)

1

UAV Historical Roots

With a concept like a controlled unmanned vehicle, it is surprising to learn how long the concept has been undertaken. Granted, the early efforts certainly did not approach the technology of today's UAVs, but the overall concept was the same.

The large numbers of early UAV developments make mentioning all of them impossible, but this chapter will cover some of the more significant systems that set the stage for the sophisticated UAVs of the 21st century.

In fact, there were early UAVs that were flying before the Wright Brothers' first flight in 1903. It has been reported, for example, that aeronautical legend Samual Langley launched a steam-powered unmanned vehicle machine over the Potomac River for a one-minute flight near the end of the 19th century.

The same year that the Wright Brothers made history at Kitty Hawk, German Carl Jatho pushed his unmanned pilotless biplane out to a distance of almost 200 feet.

A little-known fact is that primitive UAVs also performed during the first great war. The initial development of a gyroscopic guidance system really increased the capabilities of the machine. The famous Kettering Bug, a development by inventor extraordinaire Charles Kettering, brought forth a 40 horsepower-driven twin-wing vehicle. Its effectiveness was such that it was put in mass production in Dayton, Ohio. There was even a remote version of a later version of the Wright Flyer, called the Doodle Bug, that performed a drone mission.

It has been called the first UAV. The Kettering Bug is shown here in production. It served during World War I for the Allies. (USAF Photo)

This Curtis N2C-2 bi-wing aircraft was remotely controlled in research between the wars. (Curtis Photo)

Primitive UAV efforts took place in the 1930s when the Curtiss company modified one of its bi-wing N2C-2 aircraft. The RCV (Remotely Controlled Vehicle) nomenclature was used to describe the system.

During World War II, there were a number of aircraft modified to fly remotely-controlled bombing missions. One such aircraft was the PB4Y bomber that flew without a pilot in the command seat, or any other crew, for that matter.

Following the war, there are reports of a Bell P-59 fighter being modified to accomplish unmanned flight.

A modifed version of the B-24, the PB4Y model shown here was used in some missions as a remote controlled, unmanned bomber. (AF Museum Photo)

Even one Bell P-59 jet fighter performed in a remotely flown research program following World War II. (USAF Photo)

In this World War II German arrangement, an FW-190 fighter was flown off the top of a Mistel 2 bomber to crash into a ground target. (USAF Photo)

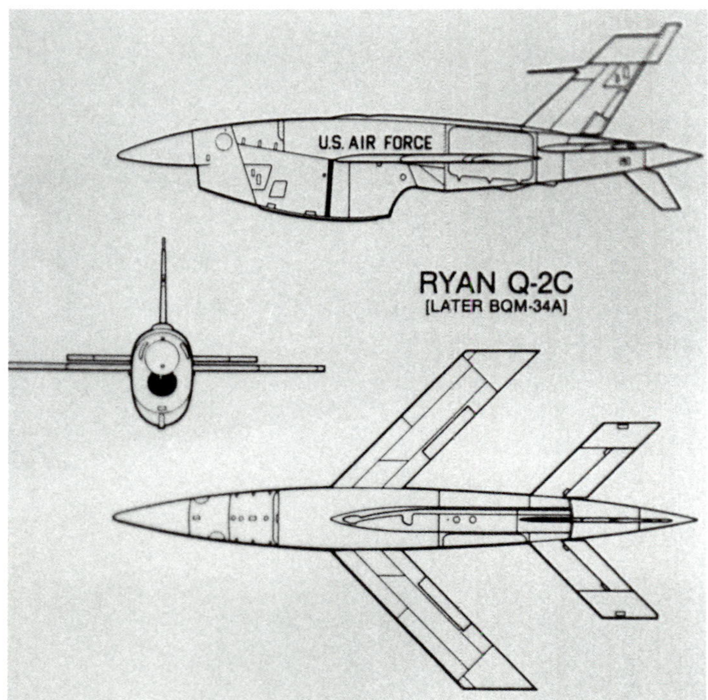

The Ryan Q-2C was one of the first pure UAVs, a system which would later evolve into the Firebee drone system. (Ryan Photo)

The U-2 high-altitude manned reconnaissance plane has been a long-time performer for USAF. In the future, some of those missions could be performed by UAVs. (USAF Photo)

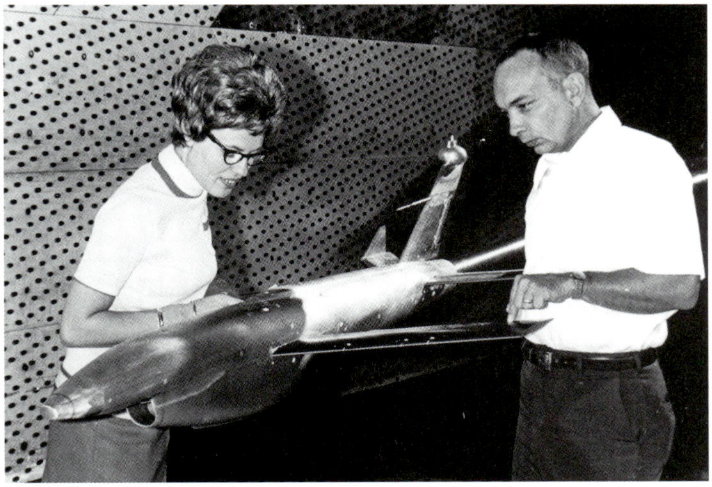

A model of the Firebee drone is tested in a wind tunnel at the Arnold Engineering Development Center (AEDC) in 1972. (USAF Photo)

This particular Firebee was the one that made the first flight for the system in 1974. (Teledyne Ryan Photo)

Most of the Firebees have been launched from airborne platforms, such as this one from under the wing of a C-130 transport. (Ryan Teledyne Photo)

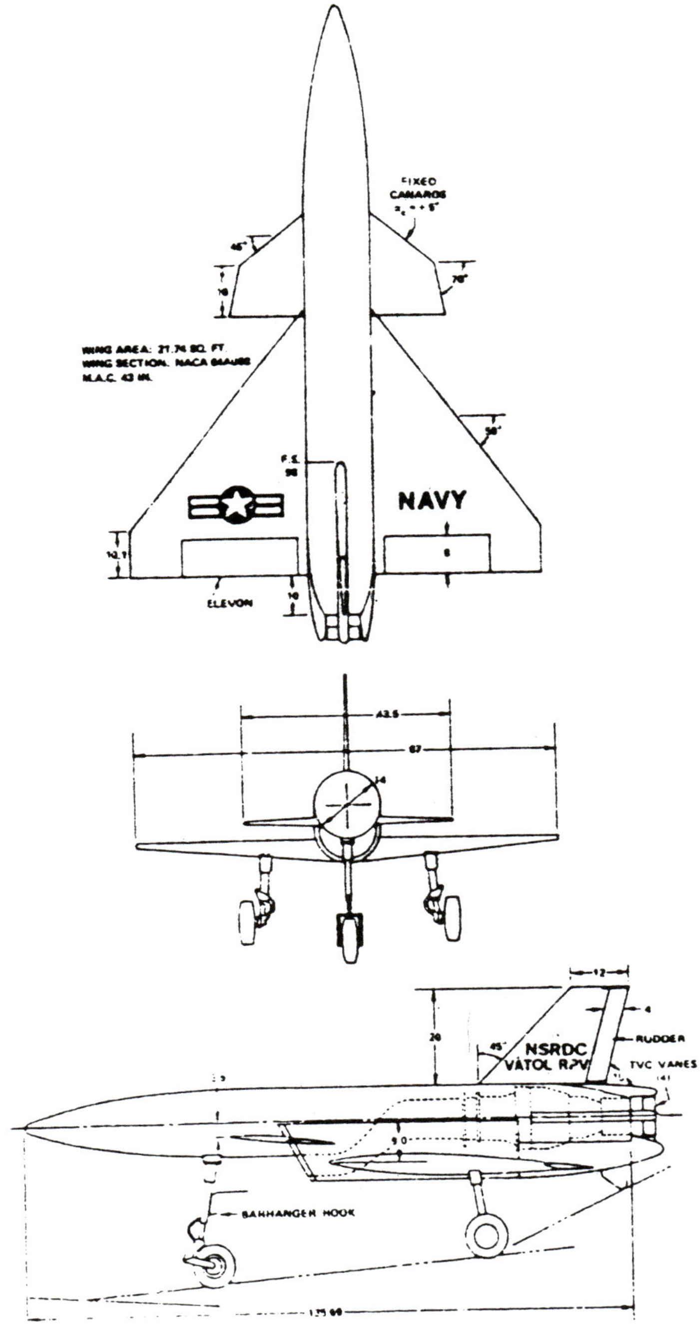

An early UAV design was this Chance-Vought delta-wing design for the U.S. Navy. (Chance Vought Drawing)

This early Lockheed UAV had a harrassment mission. It never made it to operational service. (Lockheed Photo)

The Germans used a UAV concept of sorts, but with a unique launching technique. In the arrangement, a Mistel 2 twin-engine bomber mounted an FW-190 fighter above the mother ship on a series of pylons. The fighter would then be launched off the mother ship and controlled remotely to a target. It met with only marginal success.

Stretching the application a bit, you could really define the German V-1 Buzz Bomb as a UAV. The winged vehicle was powered by a pulse-jet which, when the engine was halted, came down to earth. It flew a preprogrammed trajectory.

The payload compartment of the ARPV is shown with its covers opened. (Northrop Photo)

And then there was the interesting "Radio Control Aerial Target (RCAT)" that was also developed during the same period. Powered by a piston engine and driving a wooden propeller, the RCAT was boosted into the air by a rocket booster. The system was bought in huge numbers by the U.S. Army as the OQ-19.

The reconnaissance mission for military missions was basically a manned mission through World War II and after. The mission presented great danger to the pilot, with the capture of pilot Gary Powers after a 1960 shootdown of a U-2 high altitude spy aircraft over the Soviet Union. It was during

Northrop designed this Advanced Remotely Piloted Vehicle (ARPV) design. It was powered by a J-85 turbojet engine with a 15-foot wingspan. (Northrop Photo)

The E-45 RPV featured twin beams with uptilted wings and a piston-power pusher propulsion system. (USAF Photo)

The engine for the Beech TEDS remote vehicle was carried underneath the fuselage. (Beech Photo)

this period that unmanned systems to perform that mission were considered more seriously. To that end, a project was undertaken to modify Ryan Firebee 1 drones to fly pre-programmed reconnaissance missions.

Initially known as Model 147 and Q-2C, the new systems would serve as the forerunners of the advanced AQM-34 Firebees, and were immediately put in harm's way flying missions over China during the 1960s. A number of the systems would be shot down, and would serve as the basis of initial Chinese UAVs, when the American systems were reverse-engineered. What evolved was the initial Chinese CH-1 system.

Israel got into the UAV business in the late 1960s, initially with unpowered Maxi decoys, followed by the SAMSON system. These early developments would lead to many future Israeli sytems in the modern era, making the country one of the top producers of unmanned systems.

The U.S. Navy entered the UAV arena in the mid-1960s with the MQM-74A Chukar 1 system, with some 1,800 being built. It would be followed by an advanced C version in the 1970s. The systems were used for reconnaissance and electronic warfare missions. Italy used the Chukar as a basis for its Meteor Gufo design. Later versions of the Chukar are still an active system with the U.S. Navy early in the 2000s.

During the same period, Chance Vought designed an unmanned vehicle, designated the VATOL, which bore a resemblance to the delta wing F-102/106 fighters. In addition to the low-mounted delta wing, the early UAV also mounted a mid-fuselage front canard. The program never reached the prototype stage.

The Northrop TEDS vehicle featured a straight wing with the turbojet engine mounted underneath the rear fuselage. (Northrop Photo)

The Boeing/Compass Cope program was carried out during the mid-1970s. This photo shows the UAV making its first flight in 1974. (Boeing Photo)

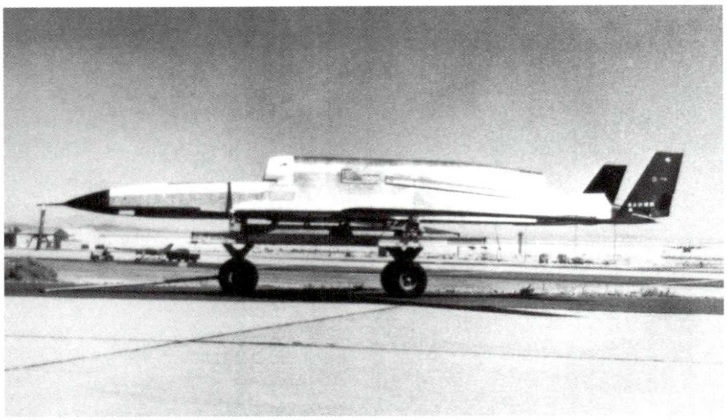

The Compass Arrow remote vehicle was similar in appearance to the Compass Cope, except for a cleaner engine mounting on the top fuselage. (USAF Photo)

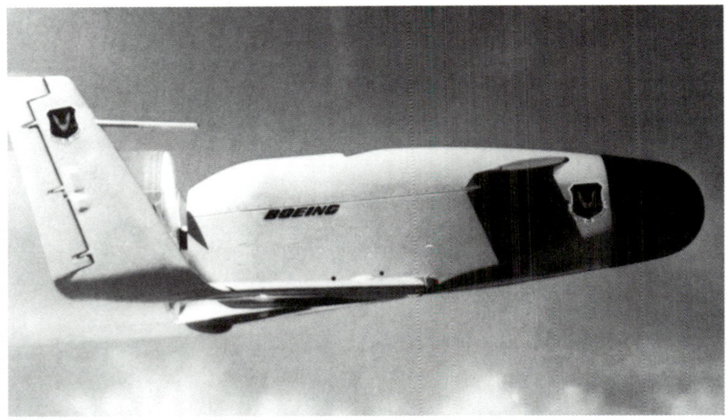

Another UAV of the time period was the Boeing YQM-121A Pave Tiger. The remotely-controlled system featured forward control canards and a pusher propeller. (Boeing Photo)

Lockheed also developed a so-called "Harrassment RDV" during the time period. The vehicle featured twin booms and a pusher propulsion system. It was launched from a ground launcher that resembled a boat trailor with a tilted launching rail.

During the 1960s, the U.S. Army developed several small propeller-driven surveillance drones. The systems were identified as the SD-1 and SD-2, but were not continued after testing. Tested from 1965 through 1977, the U.S. Army also looked at the Lockheed mini-UAV Aquila, but it also was canceled before deployment.

A 1970s UAV program was the MQM-107 Streaker, which was used by both the U.S. Army and USAF, with both services using the system during the 1980s. The Streaker looked like a manned aircraft with an underslung jet engine, along with a solid-propellant booster for launch. The system used both radio-controlled and preprogrammed guidance techniques. A number of countries bought advanced versions of the system in the 1990s.

Northrop developed its Advanced Remotely Piloted Vehicle (ARPV) in the mid-1970s for the USAF. The system was 30 feet long with a 15 foot wingspan. Powered by a J-85 turbojet engine, the ARPV was designed for reconnaissance, electronic warfare, and strike missions. It would be canceled before production.

The USAF Compass Cope-R was a high-altitude, long-range UAV designed for long duration reconnaissance. It took off and landed on conventional runways. Only two of this promising system were built. One set a world endurance record in 1974 of 28 hours and 11 minutes for unmanned, unrefueled flight.

The Compass Arrow was designed for high-altitude, low-detectability reconnaissance. It flew at 78,000 feet and incorporated many early stealth techniques. With a self-navigating Doppler Inertial guidance system and on-board computer, the UAV could fly a 2,000 mile mission.

The Philco company was involved with several small "RPV" programs, which looked more like flying model air-

This small Philco RPV (Remotely Programmed Vehicle) looks like a large model aircraft. Note the rear pusher engine system mounted above the fuselage. (Philco Photo)

Another small UAV concept vehicle was this Philco design, which was researched during the early 1970s. (Philco Photo)

This Ryan XQ-2UAV featured swept wings and tail with an embedded turbojet propulsion system. (Ryan Photo)

This Northrop flying wing remote vehicle was flown successfully in 1976. (Northrop Photo)

planes than military vehicles. One of the models featured a conventional front prop configuration, while the second carried a high-mounted fuselage pusher powerplant.

In 1976, the first Ryan Model 259 Multi-Mission UAV for USAF was rolled out. It used three interchangeable noses for different missions. The Ryan Firebrand Mach 2, ramjet-powered system served as an anti-ship missile target. After several years of successful development, the program was canceled because of budgetary reasons. It did, however, contribute valuable technology for follow-on systems.

In 1976, Northrop's "Flying Wing" remote flying vehicle made its first flight. It was followed by a similarly-appearing mini-remote piloted vehicle, known as the Very Low Cost Expendable Harassment Vehicle (VLCEHV). The new mini-

This Northrop flying wing system was identified as the Very Low Cost Expendable Harrassment Vehicle (VLDEHV) and flew in the mid-1970s. (Northrop Photo)

Left: Wind tunnel testing of the Firebrand remote vehicle. (Ryan Photo)

An F-4 fighter is shown carrying the Firebolt UAV system on its underside. (Ryan Photo)

RPVs were designed to prove out the concept of large numbers of such vehicles being used to harass enemy radar defenses. The concept was pioneered by Northrop, with a series of propeller-driven and jet-engined unmanned systems shortly after World War II.

The Firebolt was one of the most exotic UAVs built during the 1980s. A target developed for the Air Force, the system was capable of Mach 4 speeds and altitudes of over 100,000 feet. Powered by a hydrid rocket propulsion system, the Firebolt set a world altitude and speed record for unmanned sustained level flight in 1984.

The Beechcraft-designed BQM-126 was acquired by the U.S. Navy during the early 1980s as a UAV with a surveillance mission. The versatile system was capable of being launched from land, sea, and air.

The Boeing Condor UAV had very much a twin-engine transport look about it. It reached a record altitude of almost 68,000 feet in 1988. (Boeing Photo)

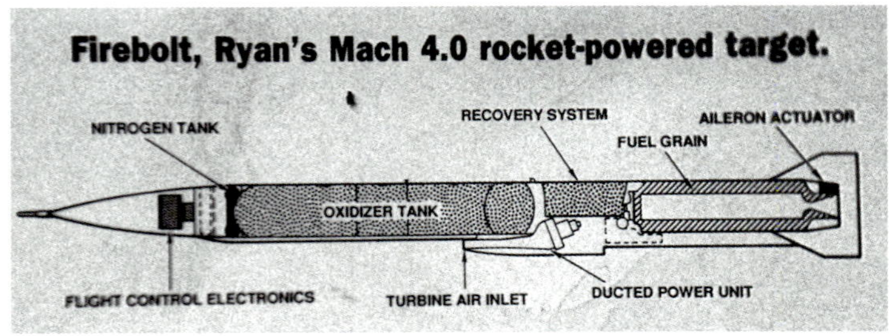

A cutaway view of the Ryan Firebolt target drone. (Ryan Photo)

An interesting UAV program that never made it to operational status was the Boeing Condor program, which made its first flight in the late 1980s. The UAV flew to an altitude of 67,000 feet and stayed on station for over two days. The program, though, was short-lived, with only eight test flights accomplished.

The Condor technology was then combined with the expertise of the famous Lockheed Skunk Works. What evolved was a stealthy UAV vehicle code-named the Q-Plane. Interestingly, the design had the provisions for a pilot should it be required. Huge dollars were spent on the program, but in 1992, the program was canceled before a single test flight was attempted.

A unique UAV design occurred with the use of the Lockheed SR-71 being the launch platform for the so-called

The D-21 remote controlled drone was carried atop an SR-71 carrier aircraft. (Lockheed Photo)

This F-86 was converted to a remote controlled target drone. (North American Aviation Photo)

The GAM-72 diversionary missile was remote controlled to simulate the launching B-52. (USAF Photo)

D-21 drone, which was carried on its fuselage top. Needless to say, the SR-71 could provide a high-altitude launch, which it certainly did, providing the D-21 with a 100,000 foot capability after its propulsion system shut down. The program was managed by the Central Intelligence Agency (CIA), but an accident during the test program caused the program to be canceled.

The Teledyne 324 Scarab was designed by Teledyne Ryan for Egypt during the 1980s. Flight testing of the UAV, a mid-wing, twin-tail vehicle, was powered by a turbojet engine mounted on the rear fuselage and ground launched from an angled launcher.

Also launched during the 1980s was the long duration Model 410 UAV. The company-sponsered program had both military and non-military missions, including border patrol, drug interdiction, anti-terrorist surveillance, forest fire patrol, and search and rescue mission. It could carry up to a 300-pound payload in its 24-cubic foot equipment bay.

Through the decades, a number of UAVs have been used for targets for air-to-air missile testing. No new vehicles had to be produced to accomplish this mission, as they were already in existence. The vehicles of which we speak are obsolete aircraft that were (and are) stored at Davis Montham Air Force Base.

A number of these high-performance aircraft have been modified for the remote controlled target mission, and the practice will undoubtedly continue for many decades to follow. A number of the planes used in these one-way missions have been the F-86, F-102, F-106, and more recently, F-4s. When used in this application, the planes are given a Q designation.

During the Vietnam War, Ryan had the UAV answer with the Model 147-H vehicle and other systems, which provided a vast amount of intelligence data for the war effort. Many of them were launched from underwing mounts on C-130 transports.

An intereesting phenomena occurred during the late 1970s when the U.S. Air Force, for all practical purposes, got out of the unmanned air vehicle business. In fact, in the early 1980s, there was not a single operational unmanned system in the U.S. Air Force inventory. That would all change shortly later, but that is the story that will be told in the following chapters, as the UAV has become the dominant aerospace system of the 1990s and the decades to follow.

2

U.S. Air Force UAVs

The USAF has long had an interest in the use of Unmanned Air Vehicles with varied missions. In fact, in the late 1990s, the Air Force used UAVs for reconnaissance purposes in the Kosovo campaign.

With the reduction in defense spending, USAF interest in the less-expensive, non-manned vehicles has contined to grow. During the late 1990s, USAF joined forces with the Defense Advanced Research Projects Agency (DARPA) to explore the future use of UCAVs for suppression of enemy defenses (See Chapter 7).

The reconnaissance mission has always been one of the most dangerous, and is perilous for manned aircraft to perform. Although there has been doubt about the UAV being an effective offensive weapon, the vehicle would appear to be the ultimate for this mission.

At the start of the new millennium, the Air Force was looking favorably at several UAVs to fullfill its immediate needs, with other systems on the horizon. With the retirement of the manned Mach 3 SR-71 reconnaissance aircraft, the need for effective UAVs has been greatly intensified.

General Atomics Predator UAV

It is a strange looking machine, with a bulbous nose and a pusher prop design. Designed and produced by General Atomics Aeronautical Systems, Inc., the 2,250 pound gross weight vehicle is powered by a minimal Rotax 912 four-cylinder engine producing only 81 horsepower.

With the bulbous nose that seems to be characteristic of UAV designs, the Predator has a wingspan of almost 49 feet, a fuselage length of 27 feet, and a maximum height of 6.9 feet. It has a maximum speed of about 80 miles per hour, a ceiling altitude of 25,000 feet, and an impressive range of 2,300 miles. The fuel capacity is about 100 gallons, with a maximum payload of 450 pounds. In the late 1990s, the Predator was available for the bargain price of only $2.3M each.

Reconnaissance is the Predator's name-of-the-game, with its 450 pounds of payload consisting of a pair of electro-optical color video cameras, an infrared video camera with three telephoto lenses, and a synthetic aperture radar. The Predator could later be equipped with the advanced high-resolution Lynx synthetic aperture radar.

This machine is operational, and moves along at about a 15,000 foot altitude, controlled remotely from the ground.

As one of the first operational UAVs, the Predator logged over 11,000 hours over three and one-half years in supporting missions over the Balkans. In 1998 alone, it flew over a hundred missions.

In Kosovo, Predators were equipped with laser spotters to designate targets enabling the use of precision-guided weapons during inclement weather. NATO commanders were able to see television-quality video from the Predators less than two seconds after it was recorded. The video was then transmitted to some 35 stations around the world.

The Predator is expected to bridge the gap until the Unmanned Combat Air Vehicle (UCAV) comes on line. As such, the Predator might pick up additional duties—offensive duties, to be exact. The system might be modified to carry a small bomb load and acquire the capability to launch missiles. It has been reported that a guided air munition could be a possible future weapon application for the Predator.

The U.S. Army at one time was interested in procuring the Predator, but despite the interest expressed by Army brass, the purchase was not completed. The system had been considered to fulfill the Army's gap in battlefield intelligence at ranges between 200 and 400 kilometers.

A follow-on system to the Predator was being designed during the late 1990s, a system called the Predator B, which could fill a gap between the 65,000 feet altitude of the Global Hawk and the 25,000 feet of the Predator.

The Predator features a tricycle landing gear, bulbous nose, and drooping rear control surfaces. (USAF Photo)

Looking like a flying saucer, this was the now-canceled DarkStar UAV. (USAF Photo)

The relatively small size of the DarkStar can be gauged from the size of the personnel around it in this photo. (Boeing Photo)

The new system was to be powered by an AlliedSignal 331 turboprop engine. Assessed at a top speed of 210 knots, it could carry a payload of about 650 pounds and fly at 45,000 feet. The Predator B will carry a 20-inch ball under the chin where the sensors will be carried. The system is a forerunner of the NASA Altair UAV.

The I-Gnat UAV is a close derivative of the Predator and is also produced by General Atomics. It has been used by the Central Intelligence Agency for a number of missions.

In 2001 a completely different capability was added to the Predator with the demonstration of an offensive capability.

The Predator showed a capability normally associated with a manned aircraft, ie the launching of air-to-surface missions. But that is exactly what a Predator did on a Nevada test range. The testing involved the firing of three Hellfire missiles, normally launched from AH-64 attack helicopters. The results were outstanding, with three tank targets being hit.

The conclusion of the testing effectively changed the Predator from a reconnaissance system to an armed and effective tank killer. Later testing planned to assess the system's capability against moving targets.

Also during 2001, the Air Force ordered more Predators, raising the inventory to approximately 40.

Boeing/Lockheed-Martin RQ-3 DarkStar UAV

To describe the DarkStar UAV in one word, the descriptor "supernatural" might be the best way. The vehicle featured a stealthy flat fuselage section that rear-mounted a large upcanted wing that seemed much too large for the fuselage.

An artist's drawing of the DarkStar banking to the left. (USAF Photo)

An interesting perspective with a DarkStar landing showing the thin profile of the UAV. (USAF Photo)

From this view of the DarkStar, it looks like little green men might emerge from the vehicle. (USAF Photo)

The DarkStar was also described as looking like "a winged horseshoe crab," with a 15-foot length and a 69 foot wingspan. The vehicle was designed to fly at 50,000 feet altitude, stay on station for eight hours, and have a range of 500 miles. The vehicles were to cost in the $12M area.

It had a reconnaissance mission, showing promise in its test program, but was surprisingly canceled in early 1999.

The DarkStar program was a joint effort of Lockheed Martin and Boeing. One of the main reasons for the cancellation of the program was reportedly because of advances made to the longstanding manned U-2 reconnaissance aircraft that might keep it operational for a number of years.

With its wings removed, the DarkStar could be carried aboard a C-130 transport and was designed to take American fighting forces further out of harm's way while giving them more accurate battlefield information via three different types of sensors that transmit data in real time.

The DarkStar was powered by a small buried 1,900 pound thrust Williams-Rolls FJ44 turbofan, enabling the vehicle to lift an 8,600 pound gross payload.

Before its cancellation, the DarkStar program was managed by the Joint Endurance UAV System Program Office at Wright Patterson Air Force Base. The program was developed in conjunction with the Tier II Plus UAV, which eventually became the Global Hawk.

The DarkStar was planned to carry one of two reconnaissance payloads, either an APG-183 SAR or CA-236 EO sensors. With this capability, the DarkStar would have had an assessed capability to penetrate thick cloud layers and image up to 14,000 square miles.

The DarkStar was the result of a 15-year secret research program, with many technical experts assembled for the effort. Surprisingly, after its cancellation, not even a core staff was retained to continue the development of technology for possible follow-on programs.

The official reasons given for the cancellation of the program were the program's high cost and the changing environment of the post-Cold War time period.

Northrop Grumman RQ-4A Global Hawk UAV

Quite simply, the Global Hawk has to be considered the most promising USAF UAV to assume a longtime operational status. First, the program greatly benefitted fiscally from the cancellation of the DarkStar, with which it was designed to complement, resulting in more emphasis being directed toward its development.

With its landing gear deployed, the DarkStar took on an entirely different look.

The Global Hawk during the production process. (Northrop Grumman Photo)

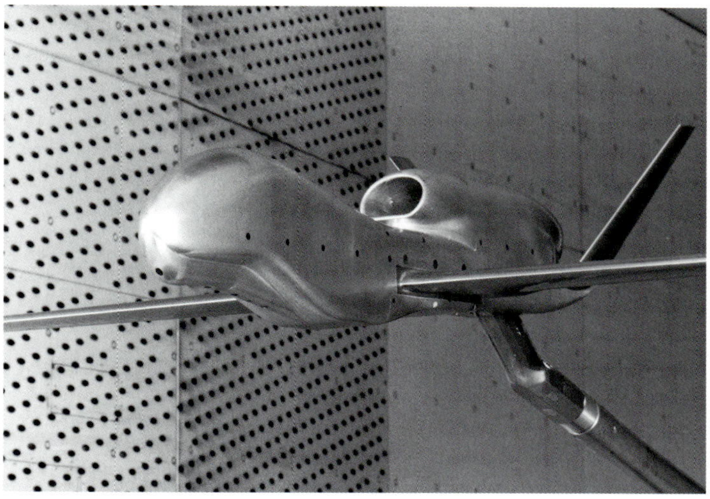

A model of the Global Hawk is tested in the 16-foot transonic wind tunnel at the Air Force's AEDC facility. (USAF Photo)

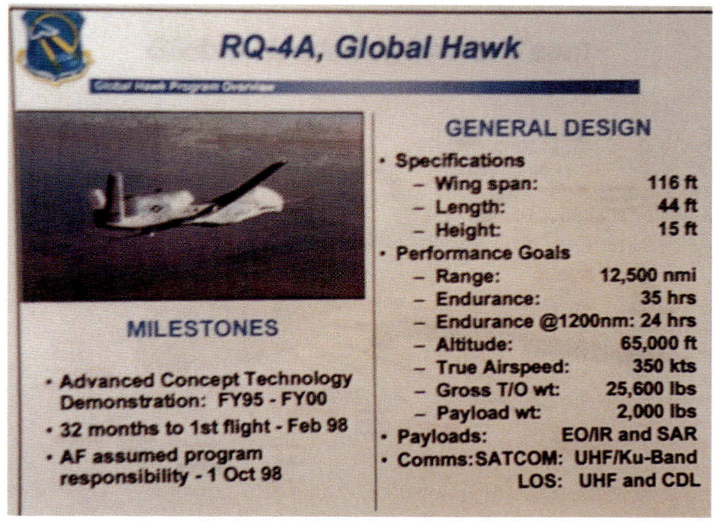

Global Hawk 2001 facts and figures. (USAF Figure)

Describing the Global Hawk is difficult, since it carries characteristics of both a manned aircraft and an unmanned drone. The fuselage looks aircraft-like, with a bulbous front section that looks like it could contain a cockpit. Its turbojet engine is mounted atop the fuselage, quite unlike any modern fighter to date. The twin tails are located in fighter locations and are quite similar to those of the F/A-18 Hornet fighter.

Then come the straight wings, certainly far different from the razor-thin swept wings of a supersonic fighter. Instead, the low-fuselage-mounted wings are perpendicular to the fuselage and stretch out on each side of the fuselage. It is easy to see that there is no supersonic flight here! A comparison might be to say the Global Hawk is closer to the A-10 Warthog than the F-15 Eagle!

Even so, the design is extremely functional for the mission it was designed to accomplish. The non-sleek machine had an advertised (1995) capability of an exceptional 14,000 mile range, and the ability to stay aloft for up to 40 hours. All this from an unmanned vehicle for the minimal price of $10 million each.

It is a much larger vehicle than the DarkStar, weighing in at almost 26,000 pounds, and powered by an Allison AE3007H jet engine rated at 7,600 pounds.

There were times during the design, by the way, that the design was considered with a new wing and the addition of a second engine.

The Global Hawk was tested with test versions. One of that number was lost in a crash when the vehicle inadvertently received a destruction command.

During a 1999 landing at Edwards Air Force Base, California, one of the Hawks veered off the runway when its nose gear collapsed. Roots for the Global Hawk stretch far back

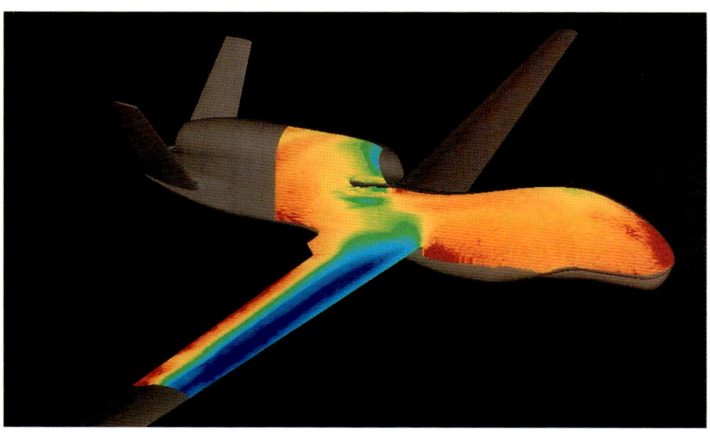

AEDC's Pressure Sensitive Pain technology gave Teledyne Ryan detailed surface pressure distribution for the Global Hawk. (USAF Photo)

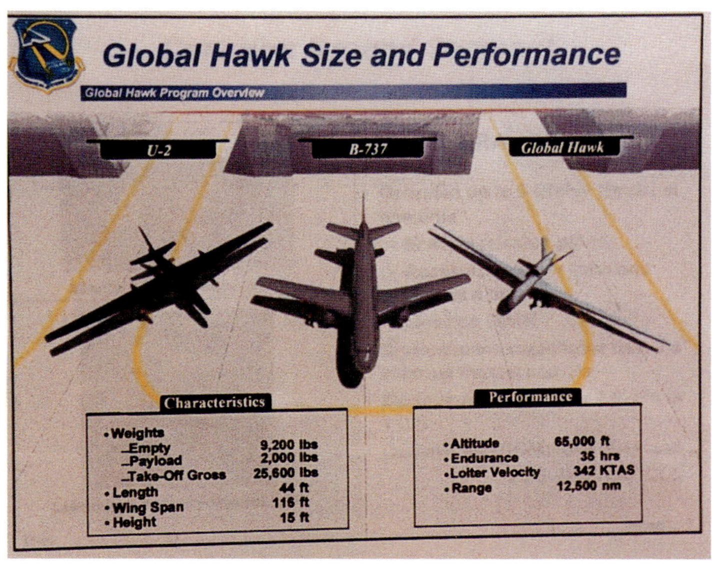

Global Hawk size and performance. (USAF Figure)

The lines and curves of the Global Hawk UAV are clear from this overhead view. (USAF Photo)

into the 1980s with classified programs that were investigating the vehicles that would replace the manned U-2 and SR-71 reconnnaissance aircraft.

The Global Hawk underwent a vigorous flight test program and demonstrated the capability to reach altitudes approaching 66,000 feet. It is assessed to be capable of surveying an area the size of the Illinois (40,000 square miles) in just 24 hours.

A ten-inch reflecting telescope provides common optics for both infrared and electro-optical sensors. The Synthetic Aperture Radar/Moving Target Indicator antenna can obtain images with three-foot resolution in its wide area search mode and one-foot resolution in its spot mode.

By about 2004, there are plans for the ASARS (Advanced Synthetic Aperture Radar System), the same unit carried by the U-2, to be incorporated into the Global Hawk. The addition could provide a coverage for an expected U-2 shortfall in the late 2000s time period.

The Global Hawk was initially developed as a long-endurance high-altitude UAV intended for multiple battlefield applications. The vehicle should provide ground commanders the ability to "see" the movements of enemy assets and enemy personnel from great distances with amazing clarity.

Like every USAF aerodynamic vehicle, the Global Hawk underwent a vigorous ground test program at the Arnold Engineering Development Center (AEDC). Models of the Global Hawk were tested in the AEDC 16-foot transonic wind tunnel during the mid-1990s.

During its flight test program, the Global Hawk flew a pair of 24-hour missions from California to Alaska and back in 1999. The flights marked the first time the UAV had flown outside the confines of the USA, as well as over water. The system's airframe is expected to last until the 2050 time period. In 2001 a Global Hawk achieved an aviation first, being the first unmanned vehicle to cross the Pacific Ocean.

Tentative plans at the start of the new millennium called for as many as 100 Global Hawks to be produced for USAF, with other countries, such as Australia, Sweden, and Israel showing interest in acquiring the system. It is also probable that advanced versions of this UAV could evolve later in the decade.

The U.S. Navy also showed interest in the Global Hawk, asking Northrop Grumman to consider the capabilities of the UAV to augment the surveillance and signals intelligence mission of the P-3 transport fleet.

Early in the 21st century it was announced that the first Global Hawk Squadron would be deployed in 2003. By 2011, the Air Force expects to have approximately one thousand airmen assigned to two or three Global Hawk squadrons.

The Global Hawk received a monumental honor in 2001, being awarded the National Aeronautic Association's 2000 Collier Trophy. The award recognized the system's performance in USAF, U.S. Navy, and NATO exercises during 2000.

Teledyne BQM-145A

This Tri-Service Medium Range UAV (MR-UAV) was chosen in 1989 to fulfill the reconnaissance mission at a relatively low cost. In the following years, it has proved its worth.

The 145A has a launch weight of one ton, has the capability to carry a 300-pound payload, is powered by a Teledyne CAE F408 turbofan, and uses a programmable guidance system. The vehicle has an overall length of about 18 feet with a wing span of 10.5 feet. It has a range of over 800 nautical miles with a top speed of Mach .91.

The BQM-145A, a Teledyne Ryan product, is a proficient UAV system for the Air Force. (Teledyne Ryan Photo)

Fighters like the F-16 could find themselves converted into UAVs in the future, as this concept has been investigated. (USAF Photo)

Even the A-10 attack aircraft has also been considered for future use in a UAV role. (USAF Photo)

Modified Unmanned Fighter/Attack Aircraft UAVs (Proposed)

Beginning in the mid-1990s, consideration was given to the modification of manned fighters and attack aircraft into UAVs. Lockheed Martin studied the concept to convert early-model F-16 fighters into long-endurance, stand-off unmanned weapons carriers. Company engineers envisioned removing the cockpit and life support systems, which would both save weight and cut drag, thus increasing range. There was also the consideration of replacing the swept wings with straight ones. With its full fuel capacity of 22,000 pounds, the aircraft could fly eight-hour combat patrols.

Lockheed indicated that the F-16 UAVs could serve in theater ballistic and cruise missile defense. The F-16s could be modified to carry up to six 2,000 pound weapons under the wings, along with a pod on the centerline station. It was predicted that three F-16 UAVs would provide the same coverage of an area that would have required a whole squadron of manned F-16s.

Also considered by the Air Force was the conversion of the potent ground-attack A-10 Thunderbolt into a UAV. It was reported that the conversion would not be that difficult. The modification would involve the addition of high-tech guidance and combat electronics, and could conceivably provide a great edge in attacking ground targets.

Also, it was indicated that as much as 1,200 pounds of life support equipment and titanium armor plating could be removed for the plane's proposed unmanned application. These concepts, though, would never produce a production system.

3

U.S. Navy/USMC UAVs

The U.S. Navy has long-felt a huge need for effective UAVs, particularly the type that can be easily employed from ship decks. To that end, the service has been looking at the possibility of vertical launch/landing systems. However, there have been a number of other systems that have been used in the interim. Some were used quite effectively in the Desert Storm conflict.

IAI/TRW Hunter UAV

This system is covered in the U.S. Army UAV chapter (Chapter 4), but it has also been used effectively by the U.S. Marines. Interestingly, Israel Aircraft Industries is one of the two prime contractors for the UAV. The Hunter was in the USJPO short-range unmanned vehicle competition in 1992, and was in low-rate production during the 1990s.

AAI/IAI Pioneer UAV

Another joint-service system was the IAI Pioneer, which was used to great success in Operation Desert Storm. In its Navy application, the system was launched from ship decks. One of its most effective missions was accurately directing fire from the 16-inch guns on Navy battleships.

Marine Corps Desert Storm commander Lt. General Boomer indicated that the Pioneer was "the single most important intelligence collector" during the operation (See Chapter 4 for system description).

Northrop Grumman BQM-74E Target UAV

One of the longstanding Navy UAV systems is the BQM-74 target UAV drone. The system has been used for a majority of Navy testing, with more than 7,400 of the system bought since 1966. The system has continued to evolve through the years with a number of improved models.

The latest version of this system is the BQM-74E, of which the Navy made a sizable purchase in 2000. The system has a 13 foot fuselage length with a 5.6 foot wing span. It is powered by a 240-pound thrust turbojet engine. Radio-controlled, the BQM-74E has a maximum speed of 610 miles per hour and a 1.7 hour loiter time. The international version is called the Chukar 3.

Even with the highly capable BQM-74E, the Navy is researching the development of a new target drone to simulate supersonic antiship missiles. Both the Navy and Air Force may replace the BQM-74, along with the BQM-34, in the mid-2000s time period.

The Hunter UAV is used by several U.S. military services. (IAI Photo)

The Pioneer UAV, which was used during Desert Storm, is shown in a dual launch from a ship deck. (U.S. Navy Photo)

The Pioneer certainly isn't built for speed, as can be discerned from this photo, but it is perfectly built for its design mission. (U.S. Navy Photo)

The Grumman BQM-74E target drone has been a mainstay for U.S. Navy testing for many years. (Grumman Photo)

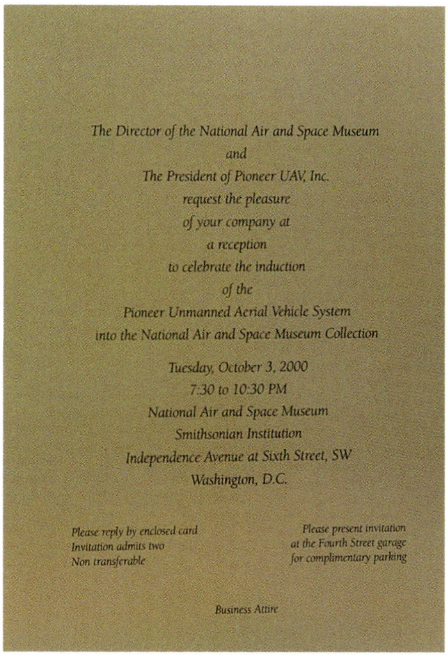

The Director of the National Air and Space Museum
and
The President of Pioneer UAV, Inc.
request the pleasure
of your company at
a reception
to celebrate the induction
of the
Pioneer Unmanned Aerial Vehicle System
into the National Air and Space Museum Collection

Tuesday, October 3, 2000
7:30 to 10:30 PM
National Air and Space Museum
Smithsonian Institution
Independence Avenue at Sixth Street, SW
Washington, D.C.

Please reply by enclosed card
Invitation admits two
Non transferable

Please present invitation
at the Fourth Street garage
for complimentary parking

Business Attire

The ultimate complement was paid to the Pioneer when one of its number was inducted into the National Air and Space Museum in October 2000. (NASM Invitation)

AeroVironment FQM-151A Pointer UAV

The Pointer is a unique man-portable, small sailplane-type UAV. The system is capable of being disassembled and troop-carried. In addition, the overall system consists of the ground operator's radio control and imagery receiving equipment.

With an amazingly low gross weight of only eight pounds, the Pointer is powered by a single electric motor, rated at 300 watts. Power is derived from either nickel-cadmium or lithium batteries.

The Pointer has a nine-foot wingspan with a six-foot fuselage length. The maximum speed of the this interesting 12-pound vehicle is reported to be about 43 knots. With a ceiling of almost a thousand feet, the Pointer can climb at about 600 feet per minute and has an endurance on station of 75 minutes with the lithium batteries. And recall, this system is hand-launched!

Size and shapewise, the Pointer looks much like a model airplane, but it is a U.S. Navy reconnaissance, surveillance, and enviironmental monitoring UAV.

The FQM-151A is used by both the Navy and Army in war games training. The first prototype flew in 1986, with the

This Pioneer waits at the ready for its next mission. (Pioneer Photo)

One of the smallest operational UAVs is the Pointer, a system which has a gross weight of only eight pounds. (USMC Photo)

Looking at the Guardian UAV gives one the impression that it is a non-earth creation. (Sikorsky Photo)

Right: Note the pair of rotating blades on the Sikorsky CL-327 UAV. (Sikorsky Photo)

Yes, the Bell Eagle Eye UAV looks like a manned helicopter, but there is no occupant onboard. (Bell Photo)

The Eagle Eye features swivelling engine pods on each wing tip, allowing for amazing maneuverability. (Bell Photo)

first systems procured by the Marines two years later. Some one hundred Pointers have been delivered to the U.S. military, with four of the systems used in the Gulf War.

Bombardier CL-327 Guardian UAV

The lunar-lander looking CL-327 appeared to have all the attributes the U.S. Navy desired for UAV shipboard applications.

The system, an upgrade of the company's CL-227, certainly has the desired VTOL capability, and entered production in 1996.

The system can be configured for both landbased or shipboard use. It has the capability of carrying a 200-pound payload. It can operate 120 miles from base for about three hours and travel at 70 knots. The 750# CL-327 also is highly suited from a shipboard point-of-view, with a special external skin to withstand the highly corrosive marine environment.

The CL-327 was considered a prime candidate for the Navy's Vertical Take-off and Landing Unmanned Aerial Vehicle (VTUAV) competition, but it was not selected. The competition was to provide a replacement for the Pioneer system sometime during the first decade of the new millenium.

Bell Eagle Eye UAV

With all the appearance of a manned helicopter, the Bell Eagle Eye was a strong competitor in the Navy's Vertical Take-off UAV (VTUAV) competition, but ended up a loser. The system was rigorously tested at the Yuma test rages, where it made 43 landings in three months of testing.

A fuselage-mounted jet engine drives a pair of wingtip tilt-rotors. It weighs only 2,250 pounds with an 18-foot fuselage length and a 15-foot wing span. With a 20,000 foot altitude capability, the Eagle Eye has a top 253 mile per hour cruise speed and an eight-hour loiter time.

At presstime, the future of the Eagle Eye was uncertain, but it was estimated at the time that it could fulfill emerging U.S. Coast Guard and UK Ministry of Defense requirements.

Three-view drawing of the Northrop FireScout. (Northrop Photo)

Northrop Grumman FireScout VTUAV

The winner of the aforementioned VTUAV competition was the Northrop Grumman FireScout vehicle which, quite frankly, also looks exactly like a manned helicopter. It should be noted that there was also a manned version of the FireScout which was used in testing the unmanned version. Eventually, the FireScout will be used by both the Navy and the USMC.

The Navy version will be able to operate on helicopter-capable ships, while the Marine Corps version will operate from a High Mobility Multi-Purpose Wheeled Vehicle. With each set-up, though, there are the same three air vehicles, a ground control station, and support equipment.

Sikorsky (Cypher II) Dragon Warrior UAV

At the end of the 20th century, the USMC stepped up with a 21st century UAV, the Dragon Warrior. Originally called the Cypher II, which was born from the earlier Cypher I system, this system is unique in that it can fly with or without wings. The Warrior weighs in at 240 pounds with a sensor payload weight up to 35 pounds. The shrouded rotor has a diameter of 3.25 feet. Its maximum speed is 125 knots, with an operational radius of 100 nautical miles. Time on station is about two hours.

The removable wings allow the vehicle to operate as a fixed wing or a rotary-wing vehicle. As a fixed-wing aircraft, the Dragon Warrior can operate in confined areas and support urban operations.

The company reported that the shrouded rotor reduces the hazard of exposed high-speed rotor blades to ground personnel. It also reduces the chance of crashing, because the rotor is protected should the aircraft nudge against a solid object.

In the early 2000s the Marines were testing a five-pound mini-UAV called the Dragon Eye. The system has a 45-inch wingspan, and can be broken down and stowed in a soldier's backpack. Amazingly, the Dragon Eye is launched by a bungee cord, and is controlled with a sending unit carried on the launching soldier's vest.

The Northrop FireScout was the winner of a recent competition and will probably become operational with the U.S. Navy. (Northrop Photo)

Sikorsky DRAGON WARRIOR

Unmanned Air Vehicle

The Sikorsky Dragon Warrior is a new UAV which can fly either with, or without, its wings deployed. (Sikorsky Photo)

Possessing a speed capability of about 45 miles per hour, an altitude capability of up to a thousand feet, and a maximum flight time of 45 minutes, the Dragon Eye was estimated to be ready for 2002 operational status, should the official go-ahead be given.

Northrop Grumman Global Hawk UAV

As is well known, the Global Hawk has been a very successful USAF program, but in 2000, the Navy also showed an interest in the system. The service looked at the Global Hawk to possibly augment its P-3 and EP-3 patrol transports. The mission would be surveillance and signals intelligence (System description in Chapter 2).

General Atomics Predator UAV

The USAF is a prime user of the Predator UAV. The system is discussed in Chapter 2, but the U.S. Navy showed continued interest in the system in the late 1990s. Many feel that a Navy version of the Predator would fulfill many of the Navy's needs.

The future status of the system with a Navy application was somewhat uncertain, but the Predator could probably at

The Dragon Warrior with its wings deployed. (Sikorsky Photo)

An artist's drawing of the Dragon Warrior with its landing gear deployed. (Sikorsky Drawing)

Like the USAF, the Navy is also looking at the possibility of converting early models, like this F/A-18A, into UAV configurations. (USMC Photo)

least serve as a testbed as the service looked at future UAV systems.

The U.S. Navy also examined the USAF Predator as a testbed for research into advanced UAV configurations. The service also looked at a modified version of the Predator for shipboard use, but rejected the proposal as being too complex.

Modified Unmanned Fighter/Attack Aircraft UAVs (Proposed)

As addressed in Chapter 2 for Air Force applications, the U.S. Navy has also shown interest in this concept, which uses modified unmanned fighters and/or attack aircraft in a UAV role. Whether this concept ever makes it to operational status with the Navy remains to be seen.

4
U.S. Army UAVs

The U.S. Army will be a prime user of UAVs in the 21st century, with a number of programs taking place in the final two decades of the 20th century. With its UAVs being launched from a ground launcher, Army UAV use is probably more flexible than with air or sea-launched versions, where logistics is more difficult. Following is a description of the prime Army UAV systems at the turn of the century:

IAI/TRW Hunter UAV

The Hunter UAV is a reconnaissance and target acquisition UAV that is based on the earlier IAI Impact Air Vehicle. It was the winner of the USJPO Short Range UAV competition in 1992, entered production in the mid-1990s, and has been an active operational system ever since.

The twin-boom airframe configuration carries a pair of Guzzi piston engines, each producing about 70 horsepower. The dimensions of the dimunitive craft are a 29 foot wingspan, 23 foot body length, and a 5.58 foot maximum height. The Hunter has a maximum weight of 1,600 pounds, an empty weight of 1,300 pounds, and a 250-pound payload weight.

Hunter UAVs use either a pre-programmed or a remote guidance technique. The UAV is recovered via a normal wheeled landing, but the capability also exists for a parachute recovery in emergency situations.

The Hunter normally carries one or two cameras, a FLIR or TV, and a laser rangefinder/designator. In addition to the Army, the Hunter has also been used by both the U.S. Navy and the U.S. Marines. The system accomplished its first operational military mission by performing reconnaissance missions in Kosovo. The system will continue to serve well into the 21st century.

In 1999, the Hunter was also a part of an Army research program linking a mixed fleet of manned aircraft and UAVs closer to being a reality. The program involved equipping a Hunter and two AH-6D Apache helicopters with the necessary communications equipment to allow the Apache pilots to control the Hunter. The Army indicated that success of the concept could increase the effectiveness of both craft and improve the survivability of both.

IAI Pioneer UAV

A highly-effective surveillance UAV, the Pioneer served with the U.S. Army, U.S. Navy, and U.S. Marines, all during Operation Desert Storm. The system was initially acquired in 1985 and was in service through the 1990s.

The Pioneer is powered by a single Sachs SF 350 piston powerplant which produces only 26 horsepower. With a payload capability of a hundred pounds and a gross weight of

First flight of the Hunter UAV on July 25, 1995. (IAI Photo)

Note the long wing span and the twin-boom configuration of the Hunter UAV. (IAI Photo)

Launch equipment for the Pioneer UAV. The UAV slides up the ramp during its early boost phase. (Pioneer Photo)

460 pounds, the Pioneer has a maximum speed capability of 95 knots, a loiter speed of 60 knots, and a maximum endurance of five and one-half hours. Its ceiling is 12,000 feet. The typical sensor package consists of a TV camera or FLIR unit.

Army use of the Pioneer during Desert Storm involved the system being launched from mobile hydraulic ramps. A typical Army Pioneer system in the field consisted of five air vehicles, a Ground Control State, a tracking Communication Unit, a Portable Control Station, four Remote Receiving Stations, pneumatic or rocket assisted launchers, and net or runway arrestment recovery system equipment. Army Pioneer operations were carried out both day and night with great effectiveness during the conflict.

There were six operational Army Pioneer units employed during the Iraqi operation, flying hundreds of combat missions. Iraqi troops in the field actually surrendered to a Pioneer UAV!

For its success, the Pioneer system was inducted into the Smithsonian National Air and Space Museum in 2000.

The Pioneer UAV in flight, showing its straight wing and tail configuration. (Pioneer Photo)

"Functional" would have to be a good descriptor of the Outrider UAV. Note the sturdy tricycle landing gear configuration. (Alliant Photo)

Power comes from the rear for the Outrider, as can be clearly seen here. (Alliant Photo)

The Outrider System

Components	Size and Weight	Performance
Four air vehicles	Wing span—3.4 m	Dash speed—110 kt
Four mission payloads —EO/IR	Overall length—3.0 m	Minimum flying speed—35 kt
Ground control equipment	Wing area—2.6 sq. m	On-station endurance
One remote video terminal	Dry weight—155 kg (340 lb)	– 4 hr for 200 km
One mobile maintenance facility per three systems	Available fuel and oil—38.6 kg (85 lb)	– 7 hr for 50 km

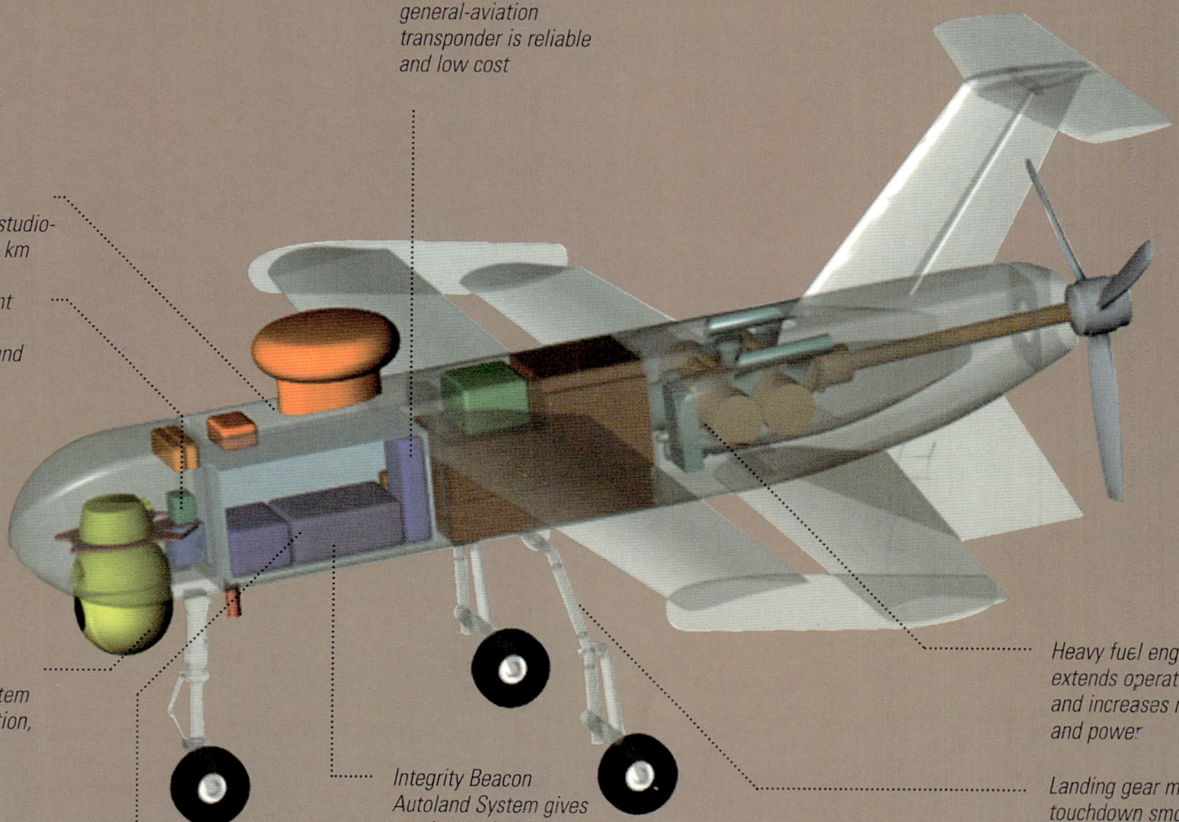

This cutaway sketch shows the internals of the Outrider UAV, including the front payload compartment, air data terminal, twin wings, and long propeller shaft. (Alliant Illustration)

Alliant Techsystems Outrider UAV

The Outrider was developed by Alliant Techsystems to accomplish a surveillance and target acquisition mission. The system was the winner against nine designs in the Joint UAV program, but would later be canceled in another competition.

The overall system consisted of four of the air vehicles, the mission payloads, ground support equipment, Global Positioning System launch and recovery system, and a remote terminal. The C-130 transport was capable of transporting the entire system. Recovery was accomplished by either wheeled landing or parachute.

The Outrider air vehicle had an interesting bi-wing configuration with an 11.1 foot span and a fuselage length of only 9.32 feet. The system weighed only 425 pounds at launch with a 50-pound maximum payload capability.

Pusher prop-power was provided by a McCulloch four-cylinder powerplant. Its maximum speed was about 110 knots with a loiter speed of 57 knots. Its ceiling was about 15,000 feet with an endurance of over four hours. Guidance is pre-programmed.

AeroVironment Pointer UAV

The Pointer is a unique man-portable small sailplane-type UAV which is used by both the U.S. Army and U.S. Navy. See Chapter 3 for Pointer system description.

An advertisement flaunting the winning of the TUAV competition by the Shadow 200. (AAI Photo)

AAI Shadow 200 UAV

One of the most important U.S. Army UAVs for the new millennium is the AAI Shadow 200 twin-boom UAV. The system assumed that position honestly, being the winner of the Army's TUAV competition against three other systems. The competition contained a fly-off during late 1999 before the announcement of the winner was made.

It was interesting that during the competition, one of the winning Shadow 200 air vehicles crashed. But the system continued to impress, meeting the Army requirements of a 31-mile range with a four-hour on station time. Another interesting requirement was that the selected UAV had to be able to run on gasoline, along with other fuels, like diesel or jet fuel. A complete unit also had to be air-transportable on two C-130 transports.

The Shadow 200's powerplant is a UAV Engines rotary engine capable of producing 37 horsepower on gasoline and 25 horses on diesel fuel. Its wing span is only 12.75 feet with a fuselage length of nine feet. Its maximum weight is 230 pounds with a maximum payload of 50 pounds.

The Shadow 200, with its rear-driven propulsion system, is shown in level flight. (AAI Photo)

The Shadow 600 is a follow-on to the 200 model. Note the tilt-back of the outer part of the wings. (AAI Photo)

The short squatty fuselage of the Hilline UAV certainly sets this vehicle apart from most other UAVs. (AeroVironment Photo)

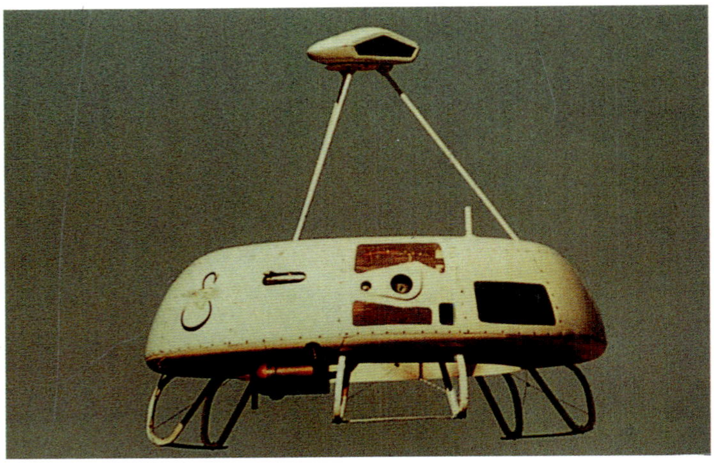

The Cypher UAV is definitely a VTOL vehicle, but a configuration that certainly could be misidentified as a flying saucer. (Sikorsky Photo)

The configuration is somewhat unique, with an inverted-V tail, twin booms, a pusher-propeller system, and an extremely slim fuselage. Undoubtedly, many different payload packages will be integrated with the Shadow 200 during the early decades of the 21st century. The follow-on Shadow 600 could also make itself felt in future years, as evidenced by its sale to a foreign customer in late 2000.

In 2001 it was announced that AAI would produce additional Shadow 200s for the U.S. Army.

AeroVironment Hilline UAV

The glider-like Hilline UAV has been used by the U.S. Army for IR signal detection, counter-narcotics surveillance, environmental monitoring, and reconnaissance missions. Powered by several different reciprocating engine types with horsepower values up to 115, the Hilline has an altitude capability of up to 50,000 feet. Its maximum speed is 109 knots.

The 800-pound UAV has a wing span of 50 feet with an overall fuselage length of 20 feet, and a maximum tail height of five feet. The UAV is capable of an unassisted take-off and can perform a conventional landing. It can remain on station for up to 24 hours.

If it looks like a helicopter, it must be the A160, which has a 300 pound payload capability (DARPA Photo)

Frontier Systems A160 UAV

During the 2000 time period, a new system called the A160 UAV was readied for flight test. Although the system was under the control of DARPA, the U.S. Army was very interested in the helicopter-style UAV. The Army envisioned a number of uses for such a system, including the extraction of troops trapped behind enemy lines.

Little data is available on the A160, but the system was thought to be powered by a commercial internal combustion engine with 300 horsepower. The system would use a rigid three-bladed propeller, with each blade being about 17 feet in length.

The payload capability of the system was expected to exceed 300 pounds, with a launch weight in the 4,000 pound range. Its operating ceiling could reach as high as 50,000 feet. The system could also evolve, in the future, to scalable larger and smaller versions of the system.

Sikorsky Cypher UAV

The Cypher VTOL UAV was used by the U.S. Army Infantry and Military Police Schools during the late 1990s. With its unique shape, the Cypher could easily be confused as a flying saucer. As such, the vehicle has no wings. Its 52 hp rotary engine is contained within the donut-shaped UAV, driving a pair of four-blade coaxial rotors, providing the Cypher with significant vertical take-off capabilities.

Performance shows a maximum altitude capability of 10,000 feet, along with a 2.5-hour on-station endurance and a maximum speed of about 74 knots. Its launch weight comes in at about 250 pounds with a 50-pound maximum payload capability.

BAI Aerosystems Tern UAV

Looking like a large radio-controlled model airplane, and not much bigger than one, the Tern has a mission of airborne chemical sensing, missle simulation, flight training, and reconnaissance.

Powered by a tiny two-stroke powerplant mounted atop the front fuselage, the Tern has a wingspan just over ten feet with a length of about eight feet. It has a maximum launch weight of only 95 pounds.

Guidance comes from a UHF uplink with a data side-band downlink, along with fiber-optic, remote control/VFR, GPS/autonavigation, and telemetry downlinks.

Suicide UAVs

The U.S. Army also has indicated an interest in so-called "Suicide UAVs." Such inexpensive systems could be used to test some of its air defense systems. The proposed systems would have a five-meter-wingspan, with a pencil-thin six-inch body diameter. With a desired price of only about $20,000 each, the UAV would have no onboard sensors.

Prophet and NBC Programs (Using UAV Systems)

During the 2000s, the U.S. Army has plans to field a modern signals intelligence, electronic attack system where UAVs would play a major function, replacing the traditional helicopter.

The Prophet program was designed to develop an airborne and ground intelligence system for division and armored cavalry regiment commanders. Several UAVs are being considered to fulfill the role.

A longer-term Army application of UAVs could be with the Army's Future Combat System (FCS), which could aid in the development of a nuclear, biological, and chemical (NBC) detection capability. Such a UAV would have to be capable of carrying a 15-pound payload of detection gear out to a 20 km range.

The engine system on the Tern looks like a bit of an afterthought, being stuck on the top of the fuselage. (BAI Photo)

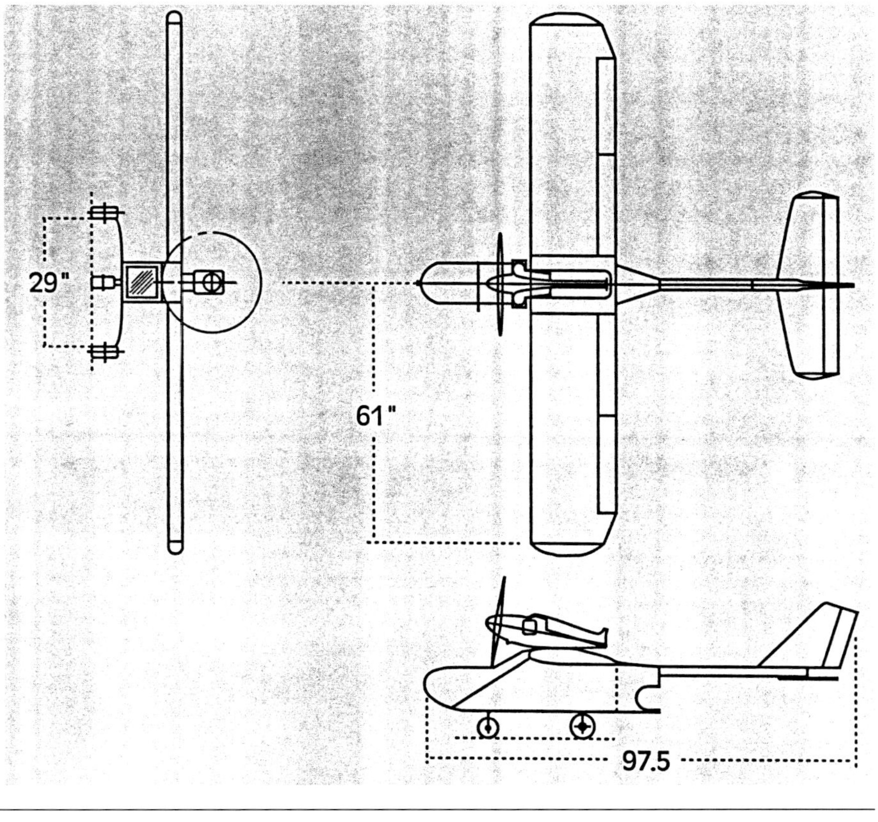

Three-view of the Tern UAV. (BAI Drawing)

5
Research and Commercial UAVs

A whole family of unique research UAVs made history with the National Aeronautics and Space Administration (NASA) during the latter part of the 20th century. Also, a number of non-military commercial UAVs have also evolved for a number of missions. A select number of these systems are addressed by this chapter.

NASA Research UAVs

During the 1990s, a family of UAVs evolved with characteristics that practically defy description. The unmanned vehicles demonstrated superlight weights, massive wingspans, large numbers of engines, and many other strange characteristics. Many of these vehicles were developed under NASA's Environmental Research and Sensors Technology (ERAST) program. There were also a number of unmanned X vehicles developed strictly to support future space programs.

Although none of these vehicles will probably ever assume an operational capability, the technology developed by this strange fleet will be a part of many future UAVs.

Altus II Dual Turbo

Appearing to look much more like a conventional aircraft than many of the other NASA research UAVs, the Altus II features a conventional-type aircraft fuselage with a low-mounted wing midway back on the fuselage. Power comes from a pusher-style four-cylinder Rotex engine equipped with a two-stage turbocharger. This propulsion system increased the power that was used in the initial Altus I configuration.

The 1,800 pound Altus II was designed to sustain flight at altitudes above 60,000 feet. Test flights during 1999 confirmed its ability to fly above 55,000 feet for some four hours. A pilot in a control station on the ground flies the craft by radio signals, using visual cues from a video camera in the nose of the Altus and information from the craft's air data system.

Apex

The Apex high-altitude research aircraft was developed by Advanced Soaring Concepts (ASC) for NASA's ERAST program. The system UAV was designed to explore the aerodyanics of controlled flight at altitudes near 100,000 feet. The launch of the vehicle was interesting, in that it was hoisted aloft tail first by a large high-altitude balloon and released at about 110,000 feet. As it gradually descends, its instrumentation collects aerodynamic data.

The remotely-piloted, semi-autonomous Apex combined a modified sailplane fuselage design with a new wing designed at MIT. The wing has a special airfoil designed for high subsonic speeds at extreme altitudes. The device extending behind the right wing is a "wake rake," which measures aerodynamic drag behind a test section of the wing,

The Altus II was flown as a performance and propulsion testbed for future high-altitude aircraft. (NASA Photo)

This computer-generated image depicts the concept of the Apex high-altitude research aircraft. (NASA Photo)

The ERAST Science UAV was based on the General Atomics Predator B system. (NASA Photo)

The Centurian remote piloted flying wing was designed to have the capability to reach altitudes up to 100,000 feet. (NASA Photo)

while a rocket pack mounted beneath the fuselage assists the Apex in transitioning to horizontal flight. Apex research flights began in 1998.

ERAST Science UAV

General Atomics Aeronautical Systems developed a modified version of its Predator B UAV, again as part of the ERAST program. The craft was to demonstrate technologies that could expand the capabilities of uninhabited aerial vehicles to perform high-altitude Earth science missions.

Centurian

Quite simply, the Centurian could be described as a solar-powered flying wing UAV. The unique remotely-piloted aircraft was also managed by the ERAST program at the Dryden Flight Research Center. The Centurian was designed and built by AeroVironment, Inc.

The Centurion was designed to reach altitudes of 90,000 to 100,000 feet to conduct atmospheric sampling and other science missions, powered entirely by solar cells which cover the wing's upper surface, providing power to its 14 electric motors, flight controls, and communications.

The Centurian is mounted on four large landing-gear platforms, each containing a pair of wheels. Initial flight testing began in 1998. Centurion served as a technology demonstrator for the follow-on Helios, a planned future high-altitude solar-powered aircraft which could fly for days or weeks at a time on science missions.

Helios

The Helios UAV made its maiden flight in 1999. Also developed by AeroVironment, the Helios has a wingspan of an amazing 247 feet and is powered, depending on configuration, by between eight and 14 motors driving broad-bladed propellers.

The prototype for the Helios solar-electric flying wing takes off on its second battery-powered check-out flight in September 1999. (NASA Photo)

The goals for this aircraft include reaching an altitude of 100,000 feet in horizontal flight and flying continuously for at least four days at altitudes above 50,000 feet. Eventual production versions of Helios will be designed to fly for six months or more on science or telecommunications relay missions at altitudes of 50,000-70,000 feet, powered entirely by the solar cells and the sun. Like the Centurian, the cells will provide all the onboard power requirements.

Raptor-Demonstrator 2

Constructed by Scaled Composites, the Raptor-Demonstrator 2 is another system in the NASA ERSAT program. The small unmanned craft features a low-wing, twin-tail configuration with a smooth upper-fuselage.

The craft has a 66-foot wing span with an overall length of 25 feet and a maximum height of 4.8 feet. Its maximum launch weight is 1,900 pounds, with a payload weight of 75-100 pounds.

The Raptor-Demonstrator 2 has accomplished eight-hour endurance and 65,000 foot altitude capabilities. (NASA Photo)

The LoFLYTE is a subscale UAV designed to aid in research for future hypersonic vehicles. (NASA Photo)

The HiMAT unmanned controlled vehicle was designed to investigate the maneuvability capabilities of future fighter aircraft. (Rockwell International Photo)

Performance-wise, the maximum altitude capability is 65,000 feet with an eight-hour endurance and a maximum speed of 80 knots. The payloads carried are atmospheric science instruments. Guidance is an autopilot with redundant hardware. It is remotely controlled by UHF uplink, along with a semi-autonomous mode.

LoFLYTE

The LoFLYTE remotely-piloted model of a hypersonic flight vehicle made its maiden, 34-second flight 93 years after the first flight of the Wright Brothers. LoFLYTE is a 100-inch-long replica of a future vehicle that might someday fly at Mach 5.5.

The vehicle was born from research in the X-30 National Aerospace Plane program. Measuring in with a 62-inch wing span and a height of only 24 inches, the LoFLYTE was constructed of fiberglass, styrofoam, and balsa wood and weighs only 80 pounds fully fueled. The LoFLYTE had a maximum speed of 250 knots.

HiMAT

The HiMat vehicle was designed to enhance transonic maneuverability of future U.S. fighter aircraft. Rockwell International developed and built two HiMATs, with first flight under NASA control occurring in 1979.

HiMat was launched from the NASA B-52 mother ship at 45,000 feet. Unmanned, HiMAT was controlled by a NASA test pilot from a ground facility that contained normal aircraft flight controls, including a throttle, stick, rudder pedals, and sensor displays. A computer converted the pilot's actions into electronic commands telemetered to the craft.

The craft was capable of speeds greater than Mach 1.5 with high-maneuverability capabilities. Its length was 22.5 feet,

The HiMAT is shown mounted in its launch position under the wing of the NASA B-52 launch aircraft. (NASA Photo)

with a wingspan of 15.6 feet and a height of 4.3 feet. The UAV was powered by a GE J85 afterburning turbojet engine, with a launch weight of 3,400 pounds.

Pathfinder

The first of the Pathfinder flying wing UAVs demonstrated use of arrays of solar cells that powered its six motors and other systems. For a time, it held the world's altitude record for propeller-driven aircraft at 71,500 feet, set in 1997.

The Pathfinder was designed and manufactured by AeroVironment and managed by NASA's Dryden Flight Research Center.

Pathfinder Plus

The giant slow-turning propellers of the Pathfinder Plus appear to be hardly moving when the majestic flying wing UAV is in flight. The remotely-piloted aircraft flies as good as it looks, setting a new world's record for solar-powered vehicles of 80,200 feet in August 1998. The altitude was also the highest ever achieved by a propeller-driven craft.

The Pathfinder Plus is a modified version of the original Pathfinder, but has a longer center wing section and eight electric motors, rather than the six on the earlier version. Both UAVs were developed by AeroVironment, Inc., as a part of the ERAST program.

Solar cells mounted on the surface of the 121 foot span wing provide power for its motors and other systems. The Pathfinder Plus was designed to serve as a slow-flying, ultra-high-altitude, long-duration aircraft that can serve as airborne an airborne platform for a wide variety of earth resources experiments and telecommunications services.

The solar-powered Pathfinder set a world record for solar-powered and propeller-driven aircraft of over 71,500 feet in a 1997 flight.

The solar-powered Pathfinder-Plus is shown during a test flight during the summer of 1998. Its maximum altitude reached was 80,200 feet, a world record for a solar-powered aircraft. (NASA Photo)

A long, slender wing and a pusher propeller characterize the Perseus B UAV. A number of different versions of the Perseus accomplished research missions during the 1990s. (NASA Photo)

The Proteus is a unique aircraft designed for high-altitude, long duration telecommunications or science missions. The craft was designed by the legendary Burt Rutan and built by the Scaled Composites company. (NASA Photo)

Perseus B

The Perseus B was the latest of three versions of the Perseus design developed by Aurora Flight Sciences. The UAV was designed to carry atmospheric sampling and other scientific or communications payloads for sustained periods at altitudes above 60,000 feet.

It is controlled by a pilot in a ground control station who monitors the aircraft insturments and a video image from an on-board camera. Among project goals were the evaluation of its high-altitude propulsion system, lightweight structures, fault-tolerant flight controls, and integration of science mission payload.

Perseus B is powered by a turbocharged four-cylinder gasoline-fueled engine driving a rear-mounted pusher propeller. It has a wingspan of 71 feet and is 27 feet long. The Perseus is designed to cruise at a speed as slow as 60 miles per hour, while carrying a payload of scientific instruments in a nose compartment. It is powered by a 105HP turbocharged Rotax piston engine. The program suffered a setback in October 1999 when the one-ton aircraft crashed on an interstate highway near Barstow, California.

Proteus

Designed by the famous Burt Rutan, the Proteus long-duration, high-altitude UAV was built for telecommunmications or science missions. It was built by Scaled Composites at its Mojave, California, development facility.

The 12,500-pound Proteus features an unconventional tandem-wing, twin-boom configuration with two rear-mounted turbofan engines providing the power. Normally flown by two

The prototype of the Theseus remotely-piloted aircraft initiates a shallow right turn during a development test flight over Rogers Dry Lake in 1996. (NASA Photo)

pilots in a pressurized cabin, the Proteus also has the potential to perform its unmanned missions flown remotely from the ground or autonomously. Rutan's design allows for Proteus to be reconfigured for differing missions, such as telecommunications relay, atmospheric research, commercial imaging, and launch of small space satellites. The aircraft features modular construction, and the ability to carry the payload either internally or externally in an under-belly pod. Removable wingtip sectons can be added or removed to tailor the Proteus' aerodyanamics for verious external payloads or for maximum altitude. The aircraft is designed to cruise at altitudes from 59,000 to more than 65,000 feet for up to 18 hours.

Theseus

The Theseus is an entirely different type of UAV, using pusher propulsion power from engines located on a pair of outboard structures which also contain the landing gears. The UAV also has a center fuselage section which mounts the craft's huge wing on its top.

Testing with the Theseus began in the mid-1990s, however, the prototype of the system was destroyed on November 12, 1996, when its right wing separated at an altitude of about 20,000 feet. The aircraft disintegrated and crashed within the Edwards test range. The failure was attributed to poor bonding between the internal structure and the skin of the all-composite 140 foot-span wing.

Theseus was designed to fly autonomously at high altitudes, with takeoff and landing under the active control of a ground-based pilot in a ground control station "cockpit."

The prototype was powered by twin turbocharged 80-hp piston engines that rotated nine-foot diameter propellers. The prime contractor, Aurora Flight Sciences, and its partners, West Virginia University and Fairmont State College, had planned for a production version of the Theseus to fly for durations of more than 24 hours at altitudes above 60,000 feet to gather atmospheric and environmental data via sensing devices mounted in three forward payload bays.

The X-33 was designed to demonstrate advanced technologies that will greatly reduce the cost of placing payloads into orbit. (NASA Photo)

X-33 Research UAV

The X-33 NASA/Lockheed Martin was a cooperative program between NASA and industry. The goal of this unmanned vehicle program was to develop a vehicle capable of making significant reductions in the cost of space delivery systems. The program was initiated in the mid-1990s.

X-34 Research UAV

With the look of a fighter aircraft, the X-34 was designed to demonstrate operational technologies applicable to future low-cost reusable launch vehicles. The X-34 was to be a suborbital vehicle launched from an L-1011 airliner and will reach altitudes up to 250,000 feet and speeds up to Mach 8.

The vehicle was designed to fly through inclement weather, land horizontally at a designated landing site, and safely abort during flight. The Orbital Sciences Corporation system has a length of 58.3 feet and a wingspan of 27.1 feet with a lift-off weight of 45,000 pounds. Propellants are LOX/RP-1 for the X-34's 60,000 pound thrust Fastrac rocket engine.

X-36 Research UAV

Looking like the next generation of a modern manned jet fighter, the X-36 UAV technolgy demonstrator shows an interesting twin-wing design and the simulation of a manned cockpit, even though no living being will fly in the vehicle.

The X-34 Reusable Launch Vehicle technology demonstrator is shown on the ramp at NASA's Dryden Flight Research Center in 1999. (NASA Photo)

The rear-wing tailless X-36 technology demonstrator research aircraft scoots across the California desert at low altitude during a 1997 research flight. (NASA Photo)

The second X-38 prototype of a Crew Return Vehicle (CRV) for the International Space Station flares for landing on the dry lakebed at Edwards Air Force Base following its fifth atmospheric test flight in March 2000. (NASA Photo)

The sub-scale vehicle was flown in the Tactical Fighter Agility Research Aircraft (TFARA) project to demonstrate critical technologies needed to improve maneuverability and survivability of future tailless, stealthy tactical military aircraft.

Designed by the McDonnell Douglas (now Boeing) Phantom Works, the small single-engine jet first flew in 1997 at Edwards Air Force Base, California. The X-36 is powered by a 700-pound thrust F112 turbofan engine with a pitch-vectoring exhaust nozzle.

About 19 feet long and with a minimal ten-foot wingspan, the X-36 used forward canards and split flaperons for maneuvering. A single-channel digital fly-by-wire control system offset built-in pitch and yaw instability. The X-36 program concluded in 1997 after 31 research flights over a six-month time period.

A pilot in a ground-station cockpit—complete with a head-up display—controlled the unmanned aircraft remotely, eliminating the need for expensive and complex autonomous flight-control systems.

X-38 Research UAV

The X-38 Crew Return Vehicle (CRV) was designed for the International Space Station. Parachute recoverable during its flight test program, the X-38 was launched from the NASA B-52 mother ship at about 40,000 feet and flew preprogrammed trajectories before its decent by parachute.

It then descended to a soft landing, similar to what the CRV lifeboat would do when returning astronatus from the space station. Later X-38 flights were planned to replicate more of the the CRV's flight operations and profile, leading to a return from orbit after being carried into orbit by a Space Shuttle in a final flight validation before the CRV enters service.

X-40A Research UAV

The X-40A is an 85 percent scale test vehicle of the experimental space plane developed for NASA by the Boeing Company. The system was designed to reduce possible risks to the larger X-37, including a drop test from a helicopter to check guidance and navigation systems for the X-37.

The unmanned system was designed to demonstrate technologies in the orbital and reentry environments for next-generation reusable launch vehicles that will increase both safety and reliability, while reducing launch costs.

X-43 Research UAV

Although the NASA X-43 hypersonic air-breathing vehicle looks quite different than the high-altitude electrical-powered vehicles, it still can be considered UAV.

The 12-foot unpiloted research vehicle was developed and built by MicroCraft Inc., while its booster was the product of Orbital Sciences Corp. The booster is attached to the aft end of the research air vehicle and accelerated to test conditions. The X-43/booster stack would be air-launched from the NASA B-52.

The X-40A test vehicle, first built for the Air Force by Boeing and successfully tested at Holloman Air Force Base, was shipped to NASA's Dryden Flight Test Center. (NASA Photo)

After launch, the air vehicle (called the Hyper-X) separates from the rocket at a predetermined altitude and velocity, and would fly a pre-programmed trajectory, conducting aerodynamic and propulsion experiments until impact into the Pacific Ocean.

Research flights were planned, two at Mach 7 and one at Mach 10. By comparison, the world's fastest plane, the SR-71, flew at slightly over Mach 3.

Commercial UAVs

Javelin Commercial UAV

The BAI Javelin is a highly versatile mini-UAV with a number of possible commercial applications. Sizewise, the glider-like configuration only weighs 15 pounds and, not surprisingly, can be hand-launched.

The system has a wing span of about eight feet and a fuselage length of six feet. Flight characteristics show a flying speed range of 20-65 mph with an altitude capability of about a thousand feet, and an on-station endurance of two hours. Amazingly, the little UAV can be turned around on the ground in only five minutes when required. Power is provided by a piston powerplant located on top of the forward fuselage.

The number of mission possibilities is practically unlimited, including the likes of live aerial video for the media, law enforcement work, and an abundance of other commercial applications.

This computer-generated illustration depicts NASA's X-43 Hyper-X hypersonic research vehicle in flight. Note the flatness of the forward fuselage. (NASA Photo)

A commercial advertisement lauding the attributes of the Javelin UAV, built by BAI Aerosystems. (BAI Advertisement)

The tiny Porter UAV is capable of performing a multitude of commercial missions. (BAI Photo)

The Model 60 UAV was designed to perform mainly maritime missions, among other commercial applications. The system is just under seven feet in length and has a span of 9.6 feet. (Freewing Photo)

Porter Commercial UAV

Developed by BAI systems and similar in appearance to a small aircraft, the Porter UAV can perform a number of commerical missions, including aerial mapping, agriculture data collection, and numerous others.

The remotely-controlled vehicle has a maximum payload capability of 70 pounds with a maximum launch weight of about 225 pounds. It has a 12-hour on-station capability with a speed capability of 65 knots. Power is provided by a 22 horsepower 12.2 cid powerplant.

Sizewise, the Porter has a wing span of 20 feet, an overall length of 12 feet, and a maximum tail height of 5.5 feet.

Launch and landing are accomplished in the conventional manner using the Porter's tricycle landing gear.

Model 60 Commercial UAV

Freewing Aerial Robotics Corp. designed and built the Model tilt-body UAV. The Model 60 can perform maritime missions or accomplish operations over obstacles into small clearings. The system could be particularly useful for the fishing industry.

A small system, the Model 60 has a wing span of only 9.6 feet, a fuselage length of 6.8 feet, and a tail height of only three feet.

Power is provided by a 12 horsepower piston engine. The launch weight is only 90 pounds, with a payload of 25 pounds.

6
Foreign UAVs

UAV development certainly is not restricted to just the United States. The rest of the world was watching, along with doing a lot on their own, through the final two decades of the 20th century.

Whether the UAV will ever realize the potential that many think it will, there certainly has been a vast amount of foreign money invested in UAV development in many different forms.

Following is a country-by-country look at how each has gone down the UAV trail with major system developments:

Argentina

The prime native Argentinian UAV was the so-called MQ-2 Bigua, which was in effect an Aussie version of the Italian Mirach-100. Development of the program began in the early 1980s.

Two versions of the Bigua have two different missions, the most important being an operational mode as a battlefield surveillance and reconnaissance version, along with a training model. Several different launching techniques have been used, including rocket boost from a ramp and later beneath a fighter aircraft.

The MQ-2 configuration sports a low-wing swept-wing design with a V-tail and a dorsel engine intake. An extremely small system, the Bigua has a wingspan of less than 12 feet. Weighing 380 kg at launch, the system can be either radio-controlled or preprogrammed. The MQ-2 is powered by a TRS 180 turbojet engine, providing a range of about 900 km.

Reportedly, during the late 1980s there was also work on a more advanced UAV, the MQ-4, but that program is believed to have been terminated.

Australia

A joint USA/Australia UAV program carries the name Aerosonde, a tiny system with a launch weight of only 30 pounds. Designed for an environmental/research mission, the Aerosonde has a wingspan of just 9.5 feet and a 5.6 foot body length. The maximum altitude is a reported 14,500 feet and a 32-hour on-station time capability.

The Jindivik UAV is a native-built UAV that has met with large success. The UAV is powered by a turbojet engine with a thousand-pound payload capability and a 4,600 pound launch weight. Through its long history, the system has been

The Jindivik of Australia features an interesting configuration where the engine system constitutes most of the vehicle. (Aerospace Technologies Photo)

Right: The Bombardier CL-327 Guardian looks something like a lunar landing vehicle, but its work is done strictly here on earth. (Bombardier Photo)

used by the U.S. Navy, along with Australian and British military forces. The country has also been involved with the USAF Global Hawk in 2001.

In addition, Australia is expected to buy a number of UAVs from foreign suppliers for its future military operational needs.

Canada

The turbojet-powered Vampire UAV, first shown in the mid-1990s. It can either be ground- or air-launched. Almost three and one-half meters in length, the UAV's engine intakes are located in the rear portion of the vehicle, with control adjusted by rear horizontal and vertical fins. Ground launch is achieved by an expendable rocket booster. With a range of over 400 km and a maxiumum altitude capability of 25,000 feet, the system was in advanced development in the mid-1980s.

The CL-89 and CL-289 are a pair of Bombardier, Inc.-built systems that were in service in 2000. The CL-89 is a 280-pound vehicle with an 8.5 foot fuselage and powered by a Williams WR2-6 turbojet engine with a solid propellant rocket booster. The CL-289 is a larger system (530 pound launch weight). Both UAVs have multiple surveillance/target acquisition/reconnaissance mission capabilities.

The CL-327 Guardian has similar mission abilities with a 770 pound launch weight. In production at the beginning of the new century, the piston-powered 750# UAV has an impressive loiter time of 6.5 hours with a maximum altitude of 18,000 feet. A VTOL design, the CL-427, which is powered by a turboshaft powerplant, will be available for service in the 21st century. The multi-mission UAV has a design loiter time of about seven hours with a top speed of 130 miles per hour.

Also, the Vindicator UAV, built by Bristol Aerospace, has been a highly-effective target UAV, having been used by the U.S. Army and Navy, along with Canadian and Saudi Arabian armed forces. The small system has a launch weight of only 180 pounds with a maximum speed of 175 knots.

China

The initial Chinese UAV systems were reportedly derived from U.S. Firebee drones that were shot down over China in the early 1990s. Those UAVs carried the Chinese designation of WZ-5.

The UAV evolution would continue until the development of the CH-1, a training/reconnaissance/civil applications system that was designed for air launch. With a 28-foot wing span and a 1,700 kg launch weight, the system uses a preprogrammed guidance system supplemented by radar tracking.

The NAI-designed CK-1C UAV was the next native effort, with the first flying prototype appearing in the mid-1970s. The system was actually an evolution of a series of unmamed fighter aircraft, with the system looking much like the Sovet La-17.

Power is provided by a MiG-19 turbojet that is suspended underneath the fuselage. The target UAV weighs about 2,400 kg with a range of up to 900 km. A command guidance system has an autopilot located in the aft fuselage which receives commands from the ground. The system was in production in the late 1980s. Little is known about a follow-on sytem, the CK-1E, which was reportedly designed to fly at low altitudes, possibly as low as 150 feet. During the 2000s,

Although the Scarab UAV was built by the U.S. Teledyne-Ryan Company, the system is operational with the Egyptian Air Force. (Teledyne-Ryan Photo)

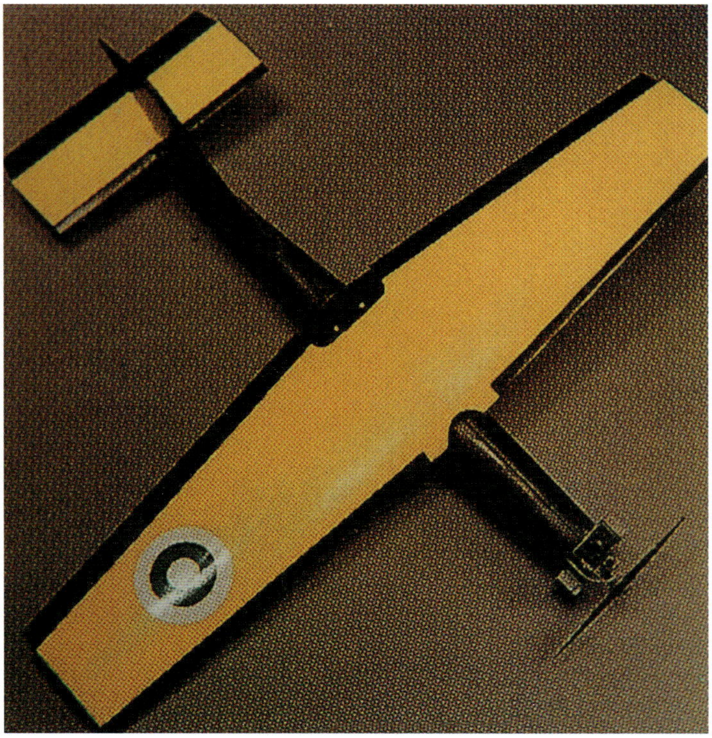

Quite frankly, the Finish AT 85 is one tiny UAV, being only 42 pounds in weight, but it has a maximum altitude capability of almost a mile.

The French Fox TS-1 is a target drone UAV powered by a piston-powered pusher engine. (CAC Photo)

As small as the French Hussard UAV is, it still is powered by a pair of piston engines, one on each wing. (Aerospacle Photo)

it is considered possible that China might attempt to consolidate stealth technology into its future UAVs.

Egypt

With a tactical reconnaissance mission, the U.S.-built Model 324 is used by the Egyptian Air Force. The 324 is launched by a solid rocket booster with parachute recovery. Powered by a thousand pound thrust turbojet engine, the system has a maximum altitude capability of 43,000 feet, a maximum speed of Mach .85, and a 2.5 hour loiter capability.

Finland

Finland has developed a pair of small target UAVs, the AT 85 and AT 97 systems. The former system is used by the Finish Navy, while the latter is used by Finish Army anti-aircraft units.

The AT 85 development started in the early 1980s, with the first production units delivered in 1985. The piston engine-powered 42 pound system was still in production in the late 1990s. With a 45 minute loiter time, the AT 85 has a maximum altitude of 5,000 feet.

The older AT 97 has roots stretching back to the late 1960s, and became operational in 1967. Smaller than the AT 85, the AT 97 weighs in at only six pounds. With a six-foot wing span, the system has the same 5,000 altitude of the AT 85 with a maximum payload of a minimal three pounds.

France

In the UAV business since the 1950s, the French started with their CT-20 performance drone for defensive missile systems. It was produced into the early 1980s, with some 1,500 constructed and several variants evolving. Powered by a turbojet propulsion system, the CT-20 weighed 660 kg at launch, had a payload capability of 150 kg, and a range of 250 km.

The CT-22 followed the CT-20, and entered active service in 1986. The drone was command-guided with a maximum range of about 1,100 km.

The ECLIPSE T1 and T2 UAVs, both rocket powered targets, are both operational training tools for the French military. The T1 has a speed capability of Mach 2.5, while the T2 can accelerate to Mach 4.3. The FOX family of UAVs includes the AT1 and AT2 versions with a reconnaissance/surveillance mission, while the TS1 and TS3 versions have target missions. All are small (12 foot wing span) vehicles, powered by 22-50 horsepower piston engines.

More recently, the Horus long-duration multi-mission UAV has evolved. The system uses the U.S. Predator airframe and 450-pound paylaods with such missions as target designation and radar jamming. In addition, there are the piston-powered Crecerelle and UGGLAN multi-mission UAVs, which have loiter times of five and eight hours, respectively.

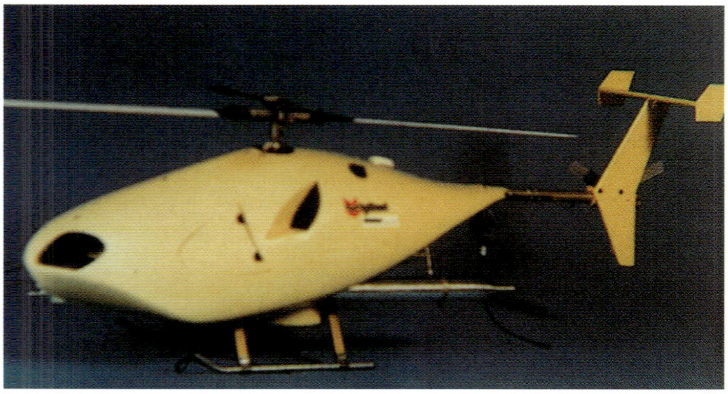

The French Vigilant UAV certainly has heavy helicopter roots. (Techno-Sud Photo)

The Heliot looks so much like a manned helicopter that you expect to see a pilot in the cockpit. (CAC Photo)

The SAE Brevel is a flat little UAV with rear-mounted square wings and a rear pusher propeller. (SAE Photo)

Other UAV systems in the large French UAV inventory are the Mart Mk2, a 240-pound UAV with a reconnaissance/surveillance mission; the Heliot (developed jointly with Italy) with a 900-pound launch weight with reconnaissance and attack missions; the Vigilant, a 1980s-designed system with muliple missions; and the Hussard system.

Then, there is also the Brevel/Tucon 95, operational with the French and German armies. The 340 pound UAV uses pusher-prop power and has a reconnaissance mission.

Germany

DaimlerChrysler Aerospace and STN Atlas Electronik are the two UAV companies in Germany. The CL-289 jet-powered reconn/surveillance/target acquisition UAV used by the German military has international implications, being deployed in Macedonia to beef up its commitment to NATO's air war against Yugoslavia. A number of the 460 mile per hour CL-289s were lost to enemy fire. In 2001, German forces completed the 1,000th launch of the CL-289.

The Seamos UAV—a VTOL system—has a maritime surveillance mission, but its tiny piston powerplant provides a top speed of only about 40 miles per hour. Carrying a payload containing radar, EO, and a flir, the Seamos can loiter over a target for about 4.5 hours at a maximum altitude of 15,000 feet.

Atlas Electronik has designed and developed a family of small UAVs with a maximum weight of 550 pounds, and with a number of different mission capabilities. The military-associated systems are all piston-powered and include the KZO, Meteo, Mucke, and Taifun. There is also a UAV effort known as Eurohawk, which is based on the USAF Global Hawk.

India

Interest in UAVs began in the 1970s in India. The first system was the PTA, which was initially tested in the mid-1980s.

Powered by a gas turbine engine suspended under the fuselage, the PTA weighs 530 kg, providing a range of 500 km, and has a Mach .74 speed capability. Production is reported to have begun in the early 1990s.

The ADE Ulka is a 360 kg target missile used for training defensive missile crews. Another recent UAV is the Nishand, which completed flight testing in the mid-1990s and could have achieved operational status in the late-1990s. The mission of the piston-powered system is assessed to be battlefield surveillance. Recovery is achieved by parachute.

Israel

Israel rates high in the research of UAV technology and the development of UAV sytems. Work on the systems is concentrated within the Israel Aircraft Industries and the Silver Arrow companies. Many of the newer Israeli UAVs have multi-mission capabilities.

The country's initial system was the Delilah, which was an unmanned decoy. Carrying the appearance of a cruise missile, the Delilah is powered by a turbojet engine and has a launch weight of about 180 kg. The system has been cleared for air launch from a number of modern U.S. fighters.

The Seamos, built by Daimler-Benz, appears similar to several Canadian VTOL systems, and has a VTOL capability. (Daimler-Benz Photo)

The Eyeview is just about as strange as it comes. The rear fuselage mounts an inverted V tail configuration. (IAI Photo)

The aforementioned Hunter UAV is produced by the Israeli IAI firm. (IAI Photo)

The IAI Firebird 2001 UAV making a perfect landing approach. (IAI Photo)

A number of piston-powered UAVs were under development in the 1990s by Israel Aircraft Industries, including the Eyeview, Hunter, and Searcher systems. These systems have reconnissance, surveillance, and target acquisition missions, with launch weights of 340, 1,600, and 820 pounds, respectively. The Firebird and Heron UAVs are reported as multi-mission systems.

Another family of military systems from Silver Arrow include: the Hermes family, with reconn/surveillance/target missions; the Micro-V and Sniper with reconn/surveillance missions; and the Calbri with a pilot-training mission.

Israel was also involved in a number of joint programs. One was the TUAV program, where it was teamed with the U.S. S-TEC and TRW companies in the development of the Sentry UAV.

The country planned to continue its UAV development in the new millennium with a high goal of locating and targeting accurately to quickly locate short-range ballistic missile mobile launchers.

The Heron is another twin-tail configuration with a rear pusher propeller. (IAI Photo)

The Searcher UAV is powered by a rear-pusher propeller and the familiar twin-boom configuration. (IAI Photo)

The Sentry is a joint U.S./Israel-developed system with a large main wing and a high horizontal tail. (TRW/IAI Photo)

Italy

The Italians have developed a family of Mirach UAVs, including the Mirach 20, 26, and 70 systems. All use piston power and have a number of different missions.

Japan

The development of the 520-pound J/AQM-1 drone was one of the first target drones from the Japanese, being developed in 1980. The program continued through the 1980s with Fuji being the prime contractor. Strictly an air-launched system, the J/AQM-1 used a preprogrammed guidance technique, but it could be overridden by either commands from the ground or the launch aircraft. A TJM-3 turbojet provides the system with a range of 350 km.

Recently, the Japanese have revealed a design effort (the RPH-2) that is a small helicopter UAV. The system was first displayed at the Paris Air Show in 1997. The piston-powered UAV has a maximum speed of 42 miles per hour with a launch weight of only 660 pounds. The system is projected to have a number of uses, including collecting agriculture data.

Fuji Industries is also responsible for the turbojet-powered family of BQM-34 target systems. Included in the family are the J and AJ KAI variants. The system weighs about 2,500 pounds and has a near sonic speed capability.

Poland

It has been reported that in the late 1990s, a combination of the Rzeszow Technical University and the PZL organization co-developed a pair of UAVs, one piston and one jet-powered, with the latter version having a greater speed of about 950 km/hr and a range of 200 km. The pusher-piston version, though, will have a much greater range.

Named the Wektor, the system features a flat-bottom fuselage, a pair of swept clipped delta wings, and a vertical tail fin. The system can be ground-launched with rocket assist or air launched.

Russia

In the early 1990s, the Sokol Design Bureau introduced the Dan target system to simulate cruise missiles and aircraft. The Dan is powered by the MD-120 turbojet, aided by rocket assist at launch. Weighing about 345kg, the system is estimated to be either preprogrammed or radio-control. The Dan entered series production in 1994.

The Mirach 20 is one of a number of systems from the Meteor CAE SPA company. (Meteor Photo)

The DR-3 is a turbojet-powered reconnaissance UAV, with both ground and air-launched versions. It is configured with rear-mounted delta-wings and small canards on the forward nose. The system is about 8.3 meters in length and weighs 1,400 kg at launch. Guidance is preprogrammed.

The M-141 (KR-17) was a jet-powered UAV developed in the late 1970s as a system with a 1,000 km range. The system achieved operational status in 1983.

The E85/E95 family of target UAVs have a target mission, with launch weights of 308 and 132 pounds, respectively. Both systems use unique pulse-jet propulsion systems, with maximum altitude capabilities of about 10,000 feet. The tactical reconnaissance R90 UAV, also pulse-jet powered, is air launched from a container.

The Ka-137 multipurpose UAV was under development in the late 1990s by the Kamov Design Bureau. Power is provided by a 65 horsepower Hirth engine. With a speed of only 95 miles per hour, the Ka-137 has an altitude capability of 16,400 feet.

The Pchela surveillance UAV entered service in 1997 (as the Stray-P). Powered by a 32 horsepower engine, the Pchela weighs about 286 pounds at launch, has a 112 miles per hour top speed, and a two-hour loiter time. The Yak-060 reconnaissaise/EW UAV started production in 1982. The 8,200 launch weight system has a 1.5 hour loiter time. It has an overall length of 8.8 feet and a 7.8 foot wingspan. The Yak-061 (Shmel) UAV, another piston-powered system, started production in 1994, is 9.1 feet in length, has a 10.6 foot wingspan, and has a 285 pound launch weight. In 2001, the Pchela was certified for night reconnaissance missions.

The Tupolev Design Bureau is the designer of a number of UAVs, all with the "Tu" identifier. The Tu-123, a strategic reconnaissance UAV, was operational during the 1980s and was powered by a pair of R-15 turbojets. With a launch weight of 84,900 pounds, the Tu-123 had a 73,000 foot altitude capability and a 1.5 hour loiter time.

The Tu-141 (VR-2) UAV, with a tactical reconnaissance mission, was powered by a single KR-17A turbojet. Possessing almost a 20,000 foot altitude capability, 152 of the model were produced, with the system serving into the late 1980s.

The Tu-243 (VR-3), another tactical reconnaissance UAV, uses a single TR-3 turbojet for power. With a launch weight of 3,100 pounds, the Tu-243 has a speed of 584 miles per hour and a 16,405 foot altitude. A Tu-300 was reportedly in development during the late 1990s to replace the Tu-243.

The NII Kulon organization is responsible for a pair of tactical reconnaissance UAVs, the Filin and Mosket (Yula) systems. The Filin, which was in development in the late 1990s, uses turbojet power with a 6,600 pound launch weight and a 20,000 foot capability. It shows a maximum speed of 590 miles per hour.

The Moskit uses piston power and has a 638 pound launch weight. With a maximum speed of 124 miles per hour, the 13 foot length system has an 11,500 foot altitude capability.

The Seeker is a product of the South African Kentron company and carries the look of many twin-boom UAV designs. (Kentron Photo)

South Africa

The South Africans introduced several native-designed UAVs in the 1990s. The Flowchart 2 and Skua target drones were designed to simulate future air threats. The Lark, a 265# UAV, has rotary engine power with an anti-jam/attack mission.

The Flowchart 2 features a flat-bottom, triangular flying wing configuration. It carries an air inlet at the mid-fuselage location. The UAV is about 15 feet in length, with a wing span of ten feet and a maximum weight of 600 kg.

The Skua was actually the basis for the Flowchart 2. The system was first tested in 1990. It has a body length slightly longer than the Flowchart, and has a payload capability of 700 kg. With its GPS guidance system, the Skua could find conversion into an air-to-surface missile.

Sweden

Sweden is looking for significant commercial applications from its CARABAS medium endurance UAV. The twin-boom fuselage system features a wingspan of 36 feet and has a projected mission for assessing forest density, along with a projected military mission of penetrating foliage for military reconnaissance.

Kentron is also the producer of the Lark UAV, a vehicle that has a configuration very similar to a delta-wing manned fighter. (Kentron Photo)

The British Sender UAV is shown in an operational ground environment.

Turkey

Developed jointly with Israel, the Ranger UAV is a piston-powered system with a 600 pound launch weight. Developed in the 1980s, the system has a reconnaissance mission. Turkey also uses USA-built Gnat 750 and I-Gnat UAVs.

United Kingdom

UAV activities during the 1990s were highlighted in Great Britain with its Phoenix system. The system received a number of continuing improvements, enabling it to perform in all weather extremes. A number of missions were planned, including use of various payloads, along with possible airborne radio relay, remote mine detection, and laser airborne targeting capabilities. The system has also proved its capability in targeting field artillery systems. It has a reported endurance time of four and one-half hours and is designed to operate at ranges out to 75 km and altitudes up to 9,000 feet.

Before the Phoenix, the British introduced the Raven UAV, a less-complex system that offered a ground surveillance capability. The system was offered to the U.S. for evaluation in 1990, but it was not accepted.

The smaller Sender UAV was designed for use at the battalion level. It could also be used by ground-based forward air controllers or artillery observers. The larger Spectator UAV was designed to support divisional reconnaissance and surveillance requirements. In the late 1990s, there were considerations for combining the two systems. Also, the Watchkeeper UAV was being pushed for a 2004 in-service date.

The Falconet is a long-term traget drone system which has design and development roots back to the 1970s, but a system that found considerable operational use in the 1990s and beyond with both the UK and France.

The UAV is about 11 feet long with a wing span of over nine feet. It can carry a payload of 50 kg, and has a maximum launch weight of 230 kg and a range of about 50 km.

The GEC-Marconi Phoenix has performed a number of military missions, including mine detection and airborne radio relay. (GEC-Marconi Photo)

In addition, the British also designed and developed the Snipe Mk5 and Banshee systems.

Finally, the UK was looking at a Tracer UAV system that was called a Combat Scout Vehicle. The decision on this system was due to be made some time in 2002.

Yugoslavia

The country has developed a solid-propellant target system called the PRM-200. It was first displayed in 1990. Its sand appearance gives the system the look of an air-to-surface missile.

Lacking any forward fins, control comes from delta-wings with rear control surfaces. The system is 2.24 meters in length with a wing span of just .88 meters. The PRM-200 has a launch weight of 90 kg and uses pre-programmed guidance.

7
A Look at Future UAVs

Entering the 21st century, the realization of the possible effectiveness of UAVs continues to be explored. A number of different concepts, with varying applications, are being examined. Some would appear to be out of this world, but a number of them will undoubtedly make it to operational status. Without a doubt, there will be many more proposals and studies of new UAVs in the first decade of the 21st century. Many, of course, will never make it beyond the paper design phase, while others might reach the prototype stage, and a miniscule percentage will make it to operational status.

A number of the more prominent efforts are covered here:

USA Future Micro UAV Concepts/Systems

Background

The UAVs that fall under this category could be the wildest, and demonstrate the highest technologies of any of the many different concepts that are being investigated. And when you say "miniature," that is exactly the case with this class, sometimes referred to as Micro Air Vehicles (MAVs).

Consider that these UAV concepts sizewise are toylike. In some cases, the complete system can actually be held in the palm of your hand! It seems like an impossible situation, but the technology is now available to support such an effort. Imagine the lack of vulnerability of such a system because of its small size!

The MicroStar is one of the number of micro UAVs which could lead to operational mini systems in the future.

DARPA is also the prime moving force in this UAV arena, with all three military branches being the possible recipients of any possible operational system that might evolve from the research efforts.

In general, the miniature UAV configurations considered bear the appearance of an advanced fighter aircraft. Other design efforts have taken the look one step further, looking more like an insect than a manmade machine.

Even with their small stature, the miniature UAVs are expected to have a unit cost as high as five figures. Making things smaller sometimes ends up costing a lot of money!

We're talking wingspans as small as a half-foot with weights measured in grams instead of pounds. Control will come from the ground, with several different techniques having been investigated. Altitude, range, and time-of-flight will be somewhat limited, but will greatly depend on the particular system.

Following are a number of the miniature UAV programs that have been investigated. Whether any of them ever make it to the field in the 21st century remains to be seen.

U.S. Army QuickLook Artillery-Launched Miniature UAV

One such application of the concept is a U.S. Army program for the development of a miniature UAV to be launched from a 155 Howitzer. The expendable system, coined the QuickLook, would cost approximately $10,000 and could provide battlefield imagery, target location, and damage assessment information.

The QuickLook vehicle is expected to be about 39 inches in fuselage length with a wingspan slightly longer. When inside the gun tube, the vehicle would appear similar to a conventional artillery round. Once clearing the end of the barrel, the system would deploy its wings and soar toward the potential target.

Once over the target, the QuickLook's propulsion system would be brought on line—possibly an internal combustion engine which might enable a 30-minute loiter time.

It is felt that such a system could provide precise target location data such that the target could be attacked in a more timely manner. Some technical problems, though, could challenge the development of such a system. It was explained that being able to communicate the wide-band communications you need for video imagery at 50 kilometers is difficult and would have to be addressed for success with this concept.

Several other launching devices were also considered, including the Crusader self-propelled howitzer, along with the possibility of mortar or rocket launch systems.

The U.S. Army was the prime investigator in this effort, with the Missile Research and Development Center at Redstone Arsenal, Alabama, and the Army Research Laboratory at Aberdeen Proving Ground, Maryland.

Lockheed/DARPA MicroStar Miniature UAV System

Another micro-UAV might be the next reconnaissance system with a vehicle small enough to fit into a soldier's pocket. It could also be smart enough to maneuver itself behind enemy lines and provide valuable intelligence information.

Amazingly, the MicroSTAR could be airborne and on its way with the flick of the wrist from the "launcher." In 1999, the system was tested during a flight test program and demonstrated flights in the 15-minute time range.

The MicroSTAR is tiny—only about three ounces in weight with a wingspan of about six inches! It is designed to collect data from an altitude of several hundred feet. The tiny micro UAV is expected to cost from $5,000-$10,000 each, stay aloft for 20 minutes, cover distances out to three miles, and cruise at speeds up to 30 miles per hour.

An onboard transmitter could provide imagery in real time as the plane is in flight. Ground commanders should be able to use the data for planning attacks. It is also possible that the model could also be used to detect chemical weapons.

AeroVironment Microbat Miniature UAV System

Then, there was the so-called Microbat ornithopter, which had the appearance of a bird, a large insect, or obviously, a bat. Its forward wings looked like they could flap, and indeed they actually could.

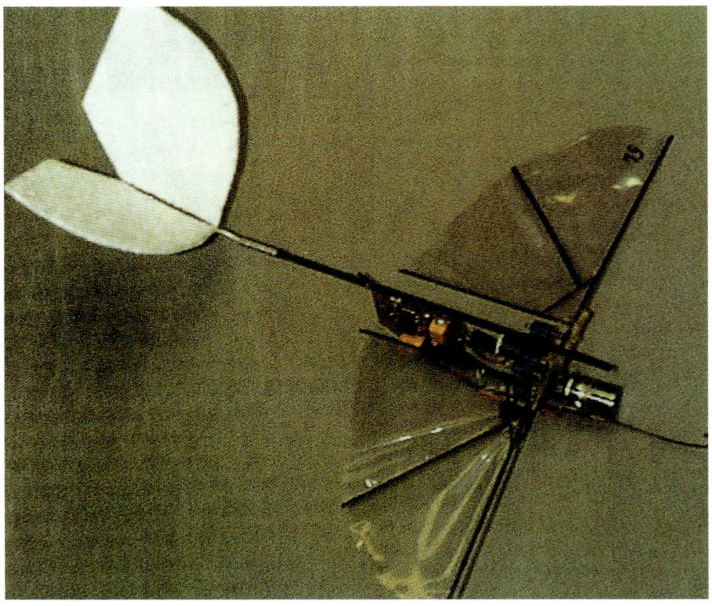

No doubt about it, the AeroVironment Microbat really does resemble an insect with its transparent wings. (AeroVironment Photo)

Built by AeroVironment, the Microbat sports a six-inch wing span and weighs only six grams. Onboard power was provided by lithium batteries. It was reported that when the batteries weakened after use, the Microbat would flutter to the ground with its wings still flopping. On some tests in the late 1990s, several sparrows landed next to the tiny craft, seeming to be looking over their "fallen comrade."

Other Miniature UAV Programs

During the latter part of the 1990s, there were also other miniature UAV programs which met with varying success.

One such program was carried out by Lockheed Martin under a DARPA contract. The program investigated the development of a miniature UAV that could carry a considerable array of payloads, including day-imaging sensors, radar jammers, or a signals intelligence or communications relay system.

The design called for a six foot long vehicle with an 85 gram fully-loaded weight. Its top speed design goal was about 30 knots, with an endurance of about 20 minutes and an altitude ceiling of 300 feet.

The craft would use an electric engine, which would make the vehicle extremely difficult to detect beyond an altitude of 100-200 feet. The individual cost of each UAV was projected to be in the $3,000-5,000 range. One worry during the design phase was the effect of operating in high winds, with restriction of such vehicles being unable to operate with wind speeds above 30 knots.

In the early 2000s, a Micro Air Vehicle with a lunar lander-type configuration was being tested. Ranging in size from six to 29 inches, the vehicle uses diesel power and a battery. The vehicle has a speed of up to 100 mph with a range from three to six miles. The estimated cost for the six-inch version was one thousand dollars each.

USA UCAV Programs

X-45 DARPA/Air Force UCAV Program

The development of an Air Force UCAV for the first decade of the new millennium was an extremely high priority goal. To that end a joint DARPA/Air Force Advanced Technology Demonstration Program addressed the development of an affordable weapon system that expanded tactical mission options and provided revolutionary new air power. The system would exploit the design and operational freedoms of relocating the pilot outdside of the vehicle to enable greater aircraft affordability while maintaining the qualities of a human operator.

In 1999, Boeing was awarded the contract to develop a demonstration system which could result in an operational system by 2010. The system has the goal of being capable of destroying enemy air defenses in a high-threat environment, in addition to carrying out certain offensive strike missions.

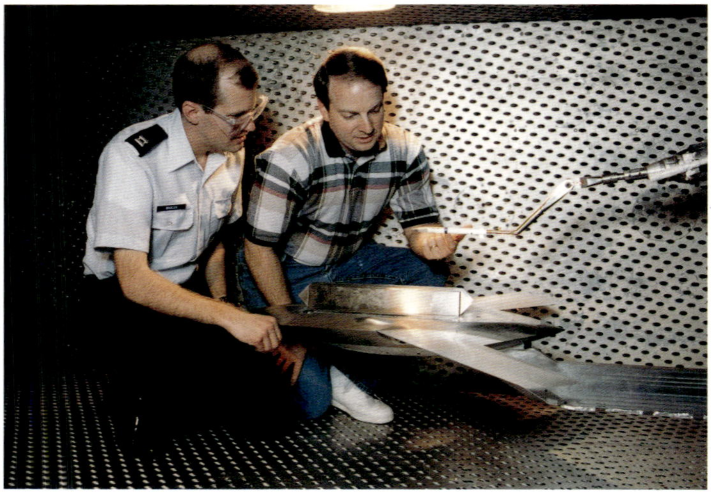

Air Force engineers examine a 1/10-scale UCAV model in AEDC's four-foot transonic wind tunnel during 2000. (AEDC Photo)

There are also considerations for using the resulting vehicle in other applications, such as peace-keeping duties.

Boeing was selected in a competition with Raytheon and Northrop Grumman for the 42-month demonstration. After the demonstration in 2005, the Air Force plans to make a decision on whether to buy the system.

The Boeing UCAV prototype, of which two were to be built, features a mid-fuselage wing with a straight leading edge and a B-2 style saw-tooth trailing edge. The wing span was expected to be about 34 feet, with power coming from an Honeywell F124 turbofan resting on the fuselage centerline. At the start of the program, though, the final decision on the engine type had not been made.

Composites will play heavily in the outer skin of the vehicle with an aluminum substructure. Boeing officials indicated that the operational version of the craft would potentially be totally composite in structure.

With a projected empty weight of about four tons, the gross weight is almost double that figure at 15,000 pounds. The craft will be very air-transportable, as the dry wings can be removed and reattached in about an hour.

The flight test program for the vehicle was planned for the 2001/2002 time period at NASA's Dryden Flight Research Center.

AIR VEHICLE

- **~15,000 / 7,500 lb Gross/Empty Weight**
- **High subsonic med/high altitude**
- **500-1000 nmi mission radius**
- **1000-3000 lb weapons payload**
- **Wide range of current & advanced weapons**
- **ESM & On-Board SAR Targeting Solution**
- **All electric**
- **Affordable Stealth to the Next Level**

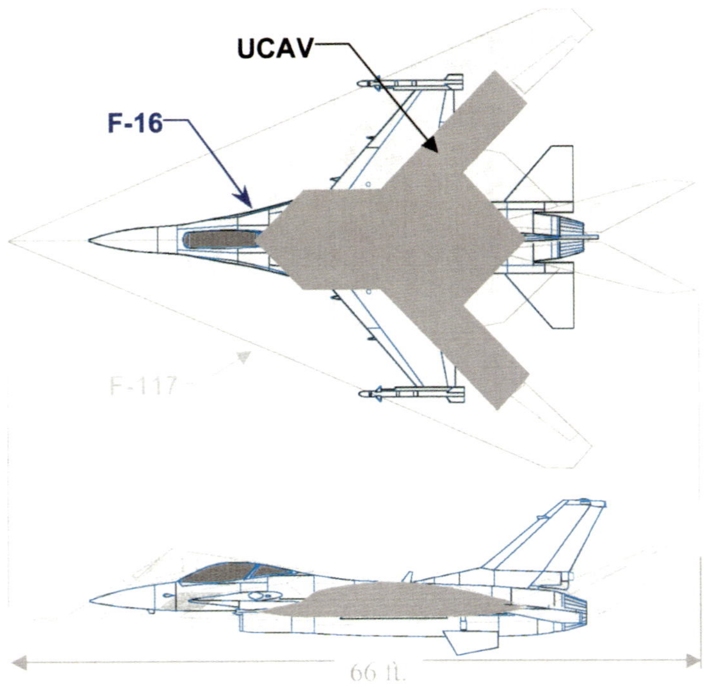

A size comparison between the USAF UCAV and an F-16 fighter. (USAF Photo)

Left: Characteristics of the Air Force UCAV. (USAF Photo)

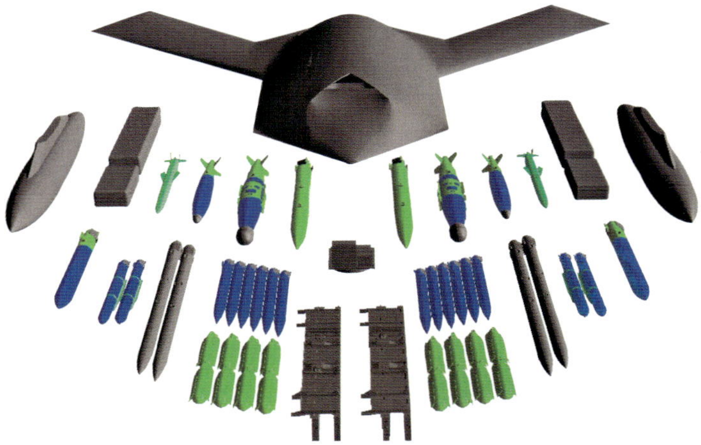

A possible weapons array for the USAF UCAV is shown in this drawing. (USAF Illustration)

A mass simulator for the USAF UCAV. Six of these containers can be transported in a single C-17 transport. (USAF Photo)

DARPA/Navy UCAV Program

In the late 1990s, the U.S. Navy stepped into the UCAV arena with a program to develop an unmanned high-performance combat aircraft which could be dominating in a wartime situation by 2030. Like the Air Force UCAV program, DARPA was also a co-sponsor with the U.S. Navy.

Boeing and Northrop-Grumman were the companies selected in 2000 to explore stealthy versions of the system, which would provide air-to-air surveillance capabilities. It was specifically stated that the systems that evolve from this design effort would not parallel the results of the Air Force study.

A high-priority goal of the new UCAV would be to acquire greater endurance than previous concepts.

The system would have the capability to survey a battlefield 500 miles away for as long as 24 hours. It could also reportedly act like like a spy satellite, relaying data to various battle groups.

The Navy UCAV was expected to be a three-to-six ton vehicle with a loiter capability of five to six hours for a strike mission. For a surveillance mission, that time could possibly be doubled.

Even before the start of this program, Northrop Grumman was considering possibilities for the UCAV. What evolved was a gull-winged vehicle which had a maximum width of 24 feet along the rear of its delta wing. The engine inlet was located high and forward on the fuselage. The weapons bay was to be located between the wheels and large enough to carry both 500 and 1,000 pound munitions. Stealth characteristics were obviously a high priority in this design with the vehicle's sharp edges and radical bends.

Naval UCAVs, by their very nature, have to be more multi-mission-oriented than their ground-based counterparts. That is because their numbers will be limited by deck space on the launching ship. Also, the Naval UCAV might tend to look more

Roll-out of the Boeing System Demonstrator UCAV Vehicle, which will be the first flight test vehicle. (Boeing Photo)

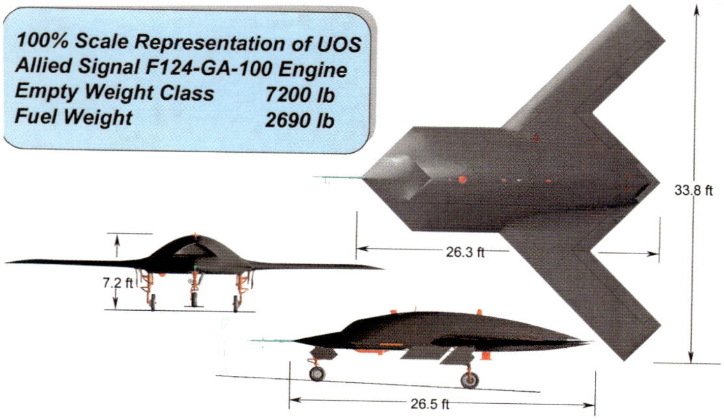

Size figures of the Boeing UCAV design. (USAF Photo)

A Northrop concept design for the Navy UCAV competition. (Northrop Drawing)

This artist's sketch shows a slightly different aft wing configuration than the previous concept. (U.S. Navy Photo)

like a manned aircraft because they will not be limited to flying only during actual combat. It is also estimated that the lifetime of a Naval UCAV might be longer than a ground-based system.

An earlier version of the Naval UCAV was designed by Northrop Grumman, and presented one of the most unique configurations ever, carrying gull-like wings which angled up, and then dropped off to the tips. The jet engine intake was located far forward on the fuselage. Finally, the low-mounted main wing was a delta design. The concept also considered the carrying of either 500 or 1,000 pound JDAM munitions.

In March 2001, the fog cleared as to the configuration of the Northrop version of the Navy UCAV. The Pegasus was certainly quite different from the earlier configurations considered. From the top view, the UCAV appears kite-like with a rear width of about 27 feet and a length of a similar dimension. The overall height is about six feet, with an airbreathing engine located on the craft's upper surface.

Air Force Sensorcraft UAV

The Sensorcraft UAV concept visualizes an unmanned vehicle with a unique square configuration. The "wings" touch the fuselage at both the front and rear of the fuselage. There are also a pair of canted rear fuselage fins, something akin to those used on the F/A-18 fighter.

As of press time, there was no firm hardware contract, just a study examining the possible mission capabilities of such a system. Along with other air vehicles, this system could

The rear view of another Navy UCAV concept. (U.S. Navy Photo)

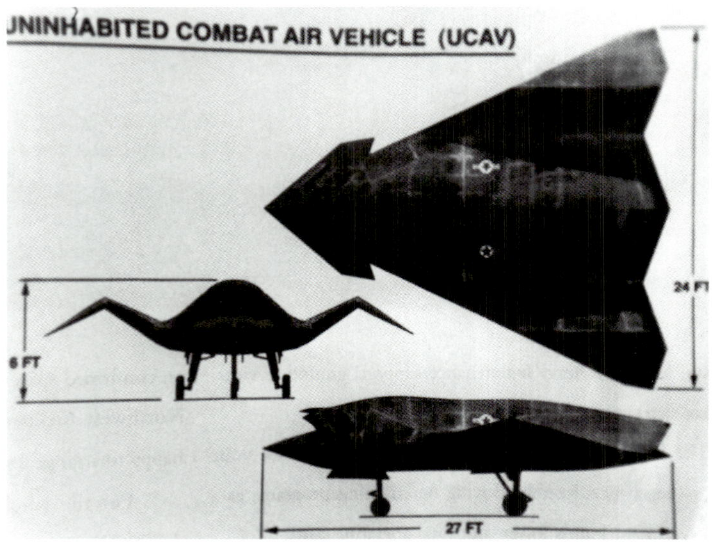

This interesting contractor UAV design shows a delta wing configuration with bird-like wings. (U.S. Navy Photo)

In 2001 the Northrop configuration of the Navy UCAV was released. With its triangular shape, it is entirely different from previous designs. (Northrop Grumman Photo)

comprise an intelligence, surveillance, and reconnaissance enterprise.

U.S. Navy Multi-Role Endurance UAV (MRE-UAV)

At the start of the new millennium, the U.S. Navy was looking at a new UAV system designed for greater range and capabilities than the current VTOL UAV systems that were under development at the time.

The new system was visualized as a medium-range system, with the hope of fielding such a system in 2005-2006. Its mission capabilities would be significant, with areas such as enemy fire suppression, command and control, and intelligence and reconnaissance.

A major mission would be the detection of short-range ballistic missiles. There might even be a capability within the MRE-UAV to carry a minimal strike capability.

Initially, the Navy had looked at a modified version of the USAF Predator, but found that shipboard operations with that system would be too difficult.

The goal of the new system would be a 24-hour on-station time with a range from the launching ship of about 200 nautical miles. Early in the program, it had not been decided whether the MRE-UAV system would utilize a VTOL or conventional takeoff/landing technique.

Boeing Dragon Fly(CRW) UAV

Another advanced UAV concept is the so-called Dragon Fly system, under development by Boeing. The system, which is directed by DARPA, uses a unique canard-rotor-wing (CRW) concept.

The design goal of the system was to be able to take off and land in a confined area without a launch or recovery system, rapidly transition to and from a fixed-wing mode, and fly at speeds in excess of 375 knots.

The 1,300 pound vehicle was designed to take off and land like a helicopter with a vertical ascent. Once in flight, the Dragon Fly would transition to a pure fixed-wing level flight, being accomplished with its top-blade, jet-powered forward flight.

For the take-off phase, jet exhaust is routed through the wingtip nozzles. By changing the pitch of the blades, the UAV can accelerate to speeds as high as 70 miles per hour. As the speed continues to increase, the jet exhaust valve begins to open, with a majority of the craft's weight now being borne by the canard surface and the tail. When a speed of about 150 miles per hour is reached, the rotor is brought to a stop in a normal wing position perpendicular to the fuselage.

Its maximum top speed is estimated to be about 375 knots. There is also the possibility that the UAV could be used to escort manned aircraft, such as the V-22 Osprey.

Whether the Dragon Fly, or some variant of the UAV, will ever reach operational status is uncertain.

Northrop Grumman Subsonic Unmanned Aircraft Proposal

The quest for the next manned bomber was being considered in the later years of the 20th century for a system that could appear in the 2030 time period. One of the designs considered was a subsonic unmanned vehicle. Three or four of these UAVs might be controlled by the third crewman in a manned aircraft. Granted, it is hard to firmly ascertain that such a vehicle might appear, but with the current interest in UAVs, it certainly can not be ruled out.

Future Foreign UAVs

(France) Dassault Unmanned Fighter UAV

It has been reported that the French Dassault Aviation Company is investigating the development of a UCAV version of

The proposed Sensorcraft has a unique configuration, to say the least. (USAF Sketch)

The rotor mounting on the CRW UAV vehicle is shown here. (Boeing Photo)

its Rafale figher aircraft. This concept would parallel similar efforts in the United States.

The system would be applicable for high-risk missions, such as suppression of enemy air defenses, strike and anti-shipping operations, and armed reconnaissance. A number of technical areas were to be first addressed, including a specific datalink for UCAV control, a large onboard processing capability, an adapted man-machine interface, and an extended data fusion process. The datalink would include a video link between the UCAV and the ground control station.

(France) Possible UCAV Development

Dassault Aviation is also examining the possibility of designing a new stealthy UCAV. This development would follow the unmanned fighter program just described.

A prototype of the system first flew in mid-2000, achieving a ten minute flight at a top speed of 250 km/hour. The jet-powered system is expected to have a maximum speed of Mach .5.

(Germany) EADS Offensive UAV/URAV Systems

The European Aeronautic Defense and Space (EADS) Company was designing, early in the 21st century, a new UAV with the capability of carrying internal offensive weapons. The design features a single-jet engine propulsion system and a twin-tail arrangement.

EADS has also been investigating a flat-fuselage URAV (Unmanned Reconnaissance Air Vehicle), a system capable of pinpointing special targets. The system is projected to have a launch weight of about 1,460 kg, with a 13.5 foot wingspan and 21-foot fuselage.

(Swedish) Saab Highly Advanced Research Configuration (Sharc) UAV

In the first years of the 20th century, the Swedish Saab Company envisioned a UAV configuration to study the effects of internal carriage of non-stealthy weapons. Sharc is designed as a complement to today's piloted aircraft, according to Saab.

The basic research for the system was accomplished in the late 1990s, with a number of wind tunnel models tested. What has evolved is a UAV configuration configured for attack missions.

The vehicle is about 32 feet in length, with a wingspan of about 26 feet, and a take-off weight of 11,000 pounds.

(United Kingdom) Future UAV Developments

In the early 2000s, the British were considering a number of new UAV systems for several new applications.

One such system employed a high-altitude, long-duration UAV as a data relay station, a role normally associated with military satellites. With a program name Extender, the system was also being examined as a relay system for communications systems. The system might also allow attack aircraft to exchange data with fighter aircraft, early warning aircraft, and tankers to provide a complete tactical picture during any operation.

In addition, the British were also considering the use of radar-equipped UAVs in tandem with the Astor Airborne Stand-off Radar aircraft. The UAV would provide more detailed surveillance data from closer to enemy lines than could be acquired by the Aster.

Index of UAVs

The following chart contains the vast majority of the UAV programs (both USA and foreign) that have evolved through the years:

USA UAV Systems

System	Nation	Power	Weight	Period	Mission
324	USA	Jet	2,400#	1990s	Reconn
A-160	USA	Piston	4,000#	2000s	Reconn
ADM-141	USA	Jet	182kg	1960s/1980s	Decoys
Aerosondi Mk1	USA	Piston	29#	1960s	Countermeasures
AF UCAV	USA	Jet	15,000#	1990s	Reconn
AQM-34L	USA	Jet	3,200#	1960s	Reconn
AQM-34V	USA	Jet	3,000#	1960s	Reconn
AQM-37	USA	Piston	620#	—	Target
AGM-37A	USA	Piston	560#	1990s	Target
AQM-37C	USA	Piston	600#	1990s	Target
AQM-37EP	USA	Piston	600#	1990s	Target
AQM-91A	USA	Jet	5,400#	1980s	Reconn
Alliance-1	USA	—	2.150#	1990s	Env/Research
Altus I	USA	Piston	1,800#	1990s	Research
Altus II	USA	Piston	1,800#	1990s	Research
Apex	USA	Piston	—	1990s	Research
Auto Blimp	USA	Piston	1,700#	1990s	Surveillance
BQM-34A	USA	Jet	2,500#	1960s	Target
BQM-34S	USA	Jet	2,500#	—	Target
BQM-74E	USA	Jet	600#	1990s	Target
BQM-126A	USA	Jet	634kg	1980s	Surveillance
BQM-145A	USA	Jet	2,000#	1990s	Target
Brave	USA	Jet	525#	1990s	Multi
Canard/Rotor/Wing	USA	Jet	1,800#	1990s	Reconn
Centurian	USA	Solar	—	1960s	Research
Chiron	USA	Piston	4,400#	1990s	Atmospheric
Combat Dawn	USA		—	1970s	Signal Intell
Compass Arrow	USA	Jet	—	1970s	Multi
Compass Cope	USA	Jet	—	1970s	Reconn
Cypher	USA	Piston	300#	1990s	Reconn
Cypher II	USA	Prop	240#	1990s	Reconn
D-21	USA	Ramjet	11,200#	1970s	Reconn
Dakota	USA	Piston	125#	1990s	Reconn/Surv
DarkStar	USA	Jet	8,600#	1990s	Reconn
Dragon Drone	USA	Piston	95#	1990s	Multi
Dragon Eye	USA	Battery	5#	2000s	Multi

Dragon Fly	USA	Jet	1,300#	2000s	Research
Eagle Eye	USA	Jet	2,000#	1990s	Reconn
ERSAT	USA	Electric	—	1990s	Research
Exdrone	USA	Piston	80#	1990s	EW, Reconn
Ferret	USA	Jet	150#	—	Multi
Firebolt	USA	Rocket	—	1980s	Target
Firebrand	USA	Ramjet	—	1970s	Target
Firescout	USA	Jet	—	1990s	Reconn
GNAT 750	USA	Piston	1,200#	—	Reconn
Global Hawk	USA	Jet	25,600#	1990s	Reconn
Harassment UAV	USA	Piston	—	1950s	Multi
Hellfox	USA	Rotary	350#	1990s	—
Helios	USA	Electric	—	1990s	Research
Hilline	USA	Piston	800#	1990s	Reconn
HiMat	USA	Jet	3400#	1980s	Research
Hunter	USA/Israel	Piston	1,600#	1990s	Reconn
Huntair	USA	Piston	425#	—	Reconn, Surv
HyFLYTE	USA	Jet	120#	1990s	Testbed
I-GNAT	USA	Piston	1,400#	1990s	Reconn
Javelin	USA	Piston/El	14#	1990s	Reconn
Kalkara	USA	Jet	1,100#	1990s	Target
Kettering Bug	USA	Piston	530#	1910s	Offensive
L-25 TUSA	USA	Jet	475#	1980s	Multi
LEARS II	USA	Piston	70#	1980s	Photo
LoFLYTE	USA	Jet	75#	1990s	Testbed
MALD	USA	Jet	90#	1990s	EW Decoy
MQM-74	USA	Jet	—	1960s	Reconn
MQM-107 (Streaker)	USA	Jet	650kg	1970s	Target
MQM-107B/D	USA	Jet	1,250	1990s	Target
MQM-107E	USA	Jet	1,100#	1990s	Target
MRE-UAV	USA	Jet	—	2000s	Multi
Mark 105 Flash	USA	Piston	200#	1980s	Research
Mark 106 Flash	USA	Piston	192#	1980s	Multi
Mi-240	USA	Piston	95#	—	Target
Micro Air Vehicle	USA	Diesel/Battery	60#	2000s	Reconn
Microbat	USA	Battery	602#	1990s	Reconn
MicroStar	USA	—	302#	1990s	Reconn
Mini-Vanguard	USA	Piston	100#	1990s	Multi
Model 60	USA	Piston	90#	1990s	Commercial
Model 100	USA	Piston	321#	1990s	Multi
Model 375	USA	Jet	350#	—	Surveillance
Model 410	USA	Jet	350#	—	Surveillance
Navy UCAV	USA	Jet	—	2000s	Research
OQ-2A	USA	Piston	110#	1940s	Target
OQ-19	USA	Piston	—	1930s	Reconn
Outrider	USA	Piston	425#	—	Surveillance
Pathfinder	USA	Electric	50#	1980s	Research
Pathfinder Plus	USA	Electric	—	1990s	Research
Pegasus	USA	Jet	—	-	2000s Multi
Penetrator	USA	Jet	907kg	1980s	Multipurpose
Perseus B	USA	Piston	2,000#	1980s	Research
Pioneer (RG-28)	USA	Piston	460#	1980s	Multi
Pointer D-2	USA	Electric	8#	1990s	Surveillance
Porter	USA	Piston	200#	1990s	Sensor Dev.
Predator(RQ-1A)	USA	Piston	2,250#	1990s	Reconn

Predator B (Altair)	USA	Turboprop	—	2000s	Reconn
Proteus	USA	Jet	12,500#	1990s	Telecomm
Prowler 2	USA	Piston	700#	1990s	Surveillance
Quick Look	USA	Gun	—	1990s	Reconn
Raptor D2	USA	Piston	1,900#	1990s	Research
SASS LITE	USA	Piston	1,900#	1990s	Multi
Scorpion 60	USA	Piston	85#	1990s	Multi
Scorpion 100	USA	Piston	—	1990s	Multi
Sea Bat	USA	Piston	200#	—	Surveillance
Sea Snake	USA	Ramjet	—	1990s	Target
Sender	USA	Electric	220#	—	Surveillance
Sensorcraft	USA	Jet	—	2000s	Multi
Sentry (STM-5B)	USA	Piston	250#	1980s	Multi
Shadow 200	USA	Rotary	300#	1990s	Reconn
Shadow 600	USA	Rotary	600#	1990s	Reconn
Skyeye R4E	USA	Rotary	1,250#	1990s	Multi
Super MOM	USA	Jet	1,440#	1990s	Target
Swallow	USA	Electric	60#	—	Surveillance
Tern	USA	Piston	95#	1980s	Multi
Theseus	USA	Piston	7,000#	1990s	Atmospheric
Truck	USA	Piston	120#	1990s	Testbed
UAV 18-50	USA	Piston	772#	—	Reconn
VATOL	USA	Jet	—	1950s	Reconn
VLCEHV	USA	Jet	—	1970s	Electronics
Vanguard	USA	Piston	235#	1990s	Multi
W570A	USA	Jet	45,000#	—	Reconn
X-33	USA	Rocket	—	1990s	Research
X-34	USA	Rocket	45,000#	1990s	Research
X-36	USA	Jet	1,300#	1990s	Research
X-38	USA	Rocket	—	1990s	Research
X-40A	USA	Rocket	—	1990s	Research
X-43 (Hyper-X)	USA	Scramjet	—	2000s	Research
X-45A	USA	Jet	15,000#	2000s	Strike

Foreign UAV Systems

System	Nation	Power	Weight	Period	Mission
ADE Ulka	India	Rocket	360KG	1990s	Target
ADS 95	Swiss	Piston	611#	1990s	Target
ALBA	Spain	Piston	50#	1990s	Target
ALD	Spain	Piston	60#	—	Observation
AT85	Finland	Piston	42#	1980s	Target
Aerosonde Mk. 1	Aust	Jet	300#	1990s	Env/Research
AVE	France	Jet	60kg	2000s	Multi
B-2	China	Piston	124#	—	Target
BQM-34	Japan	Piston	2,500#	1990s	Target
Banshee (BTT3-342)	UK	Piston	190#	1980s	Target
Banshee (BTT3-731)	UK	PIston	190#	1980s	Target
Brevel	Ger/Fr	Piston	330#	1990s	Surveillance
C-22L	France	Jet	1,388#	1980s	Target
CH-1	China	Jet	1,700kg	1970s	Target/Reconn
CK-1C	China	Jet	2,400kg	1970s	Target
CK-1E	China	Jet	——	2000s	Target
CL-89	Canada	Jet	280#	1990s	Surv/Reconn
CL-289	Can/Ger.	Jet	530#	1990s	Surv/Reconn

CL-327	Canada	Jet	750#	1990s	Surv/Reconn
CL-427	Canada	T-Shaft	750#	2000s	Surv/Reconn
CT-20	France	Jet	660KG	1950s	Traget
Cabri	Israel	—	—	1990s	Training
CARABAS	Sweden	Piston	—	1990s	Multiple
Changkong IC	China	Jet	5,400#	—	Target
Colibri	Israel	—	50#	2000s	Reconn/Surv
Crecerelle	France	Piston	300#	1990s	Reconn
D-4	China	Piston	308#	—	Target
DR.3/VR3	Russia	Jet	——	1970s	Reconn
Dan (Tribute)	Russia	Jet	345kg	1950s	Target
Dassault Fighter	France	Jet	—	1990s	Multi
Dililah	Israel	Jet	180kg	1990s	Decoy
Dragon	France	Rotary	330#	—	Research
EADS UCAV	EADS	Jet	4,500kg	2000s	Multi
E85/E90	Russia	Jet	310#	1980s	Target
E95	Russia	Jet	130#	—	Target
E-Hunter	Israel	Piston	2,100#	1990s	Multi
ECLIPSE T1	France	Rocket	300#	1990s	Target
ECLIPSE T2	France	Rocket	450#	1990s	Target
Eurohawk	Germany	Jet	26,000#	—	Reconn/Surv
Extender	UK	—	—	2000s	Comm
Eyeview	Israel	Piston	340#	1990s	Reconn
Falconet	UK	Jet	450#	1990s	Target
Filin	Russia	Piston	6,600#	2000s	Reconn
Firebird	Israel	Piston	340#	1990s	Reconn
Flowchart	S Africa	Jet	600kg	1980s	Simulation
FOX AT1/AT2	France	Piston	160/250#	1990s	Surv/Reconn
FOX T51	France	Piston	150#	1990s	Target
FOX TS3	France	Piston	220#	1990s	Target
FOX TX	France	Piston	250#	1990s	EW
Harpy	Israel	—	—	—	SEAD
Heli-25	Israel	Piston	55#	1990s	Multi
Heliot RPM	Fr/Italy	Piston	900#	1990s	Reconn/Attack
Hermes 450	Israel	Rotary	1,i00#	1990s	Reconnaissance
Hermes 1500	Israel	Rotary	3,300#	1990s	Multi
Heron	Israel	Piston	2,400#	1990s	Weather, SIGINT
Hermes	Israel	—	—	1990s	Multi
Horus	France/USA	Piston	—	1990s	Multi
Hunter	Israel	Piston	1,600#	1990s	Reconn/Surv
Hussard	France	Piston	55#	2000s	Reconn/Surv
Jindivik	Ausie	Piston	4,600#	1940s	Drone
IMP	UK	Piston	—	1990s	Training
J/AMQ-1	Japan	Jet	520#	1980s	Target
Ka-37	Russia	Piston	550#	1990s	Multiple
Ka-137	Russia	Piston	615#	2000s	Multi
Kolibri	Russia	Piston	—	1990s	Reconnaissance
La-17MM	Russia	Jet	6,740#	1950s	Target
Lakshya (PTA)	India	Jet	630KG	1970s	Target
Lark	S Africa	Rotary	265#	1990s	Anti-jam/attack
Luna	Germany	Piston	44#	—	Reconn/Target
M-23	Russia	Jet	31,400#	1980s	MiG-23 Target
M-141	Russia	Jet	—	1970s	Reconn
M-932	Russia	Jet	760#	—	Target
MQ-2 (Bigua)	Argen.	Jet	310kg	1980s	Target/Reconn

Mart Mk2	France	Piston	240#	1990s	Reconn/Surv
Marula	France	Rotary	300#	1990s	Reconn/Surv
Micro-V	Israel	Piston	110#	1990s	Reconn/Surv
Mini-Sheddon	Israel	Piston	55#	1990s	Recon/Surv
Mirach 20	Italy	Piston	375#	1980s	Surv/Target/Acq.
Mirach 26	Italy	Piston	440#	1990s	Multiple
Mirach 70	Italy	Piston	405#	1960s	Target
Mirach 100/4	Italy	Jet	620#	1980s	Target
Mirach 100/5	Italy	Jet	705#	1970s	Target
Mirach 150	Italy	Jet	760#	1990s	Reconn
Moskit	Russia	Jet	640#	1990s	Reconn
Mucke	Germany	Piston	330#	2000s	EW
Nishant	India	Piston	—	1990s	Reconn/Surv
Nitrohawk	GB	Piston	21#	1980s	Targeting
PRM-200	Yugo	Rocket	90kg	1980s	Target
PTA	India	Jet	530kg	1980s	Reconn
Pchela	Russia	Piston	—	1990s	Reconn
Pretit-Duc	France	Jet	60kg	2000s	Multi
Petrel	UK	Rocket	285#	1990s	Ballistic Target
Phantom	UK	Piston	65#	1990s	Reconn/Surv
Phela (Stroy-P)	Russia	Piston	286#	1990s	Surv
Phoenix	UK	Piston	400#	1980s	Surv/Acquisition
R90	Russia	Jet	99#	—	Tactical Reconn
RPH-2	Japan	Jet	660#	1990s	Multi
Ranger	Is/Swit	Piston	600#	1980s	Reconn
Raven	UK	Piston	190#	1990s	Reconn/Surv
RPO Midget	Sweden	Piston	100#	—	Reconn/Surv
S-Mart	France	Piston	317#	2000s	Surv/EW
SIVA	Spain	Piston	440#	1990s	Reconnaissance
Scarab	Egypt	Rocket	—	1990s	Surveillance
Seamos	Germany	T-Shaft	2,300#	1970s	Surv/Others
Searcher	Israel	T-Prop	820#	1990s	Surv/Acquisition
Seeker	S Africa	Piston	560#	1980s	Reconn/Acqui
Sender	UK	——	—	—	Photographical
Sentenel	Canada	T-Shaft	420#	1970s	Reconn/Acqui
Sharc	Sweden	Jet	11,000# 2000s	Attack	
Sheddon Mk.3	Israel	Piston	90#	1990s	Multi
Shmel	Russia	Piston	300#	1990s	Surveillance
Silver Arrow	Israel	—	—	1990s	Multi
Skua	S Africa	Jet	700kg	1980s	Target Drone
Snipe Mk5	UK	Piston	145#	1990s	Target
Snipe Mk15	UK	Piston	—	1990s	Target
Sniper	Israel	Rotary	330#	2000s	Reconn/Surv
Sojka	Chk/Hung	Piston	—	1990s	Reconnaissance
Spectre	UK	—	—	1990s	Surv/EW
Spectator	UK	—	—	2000s	Surveillance
Sperwer	France	Piston	400#	1990s	Reconn/Acqui
Sprite	—	—	—	—	—
Tacun	Germany	Piston	330#	—	Reconn/Surv
Taifun	Germany	Piston	300#	2000s	Reconn/Attack
Tu-123 (Yastreb)	Russia	Jet	84,700#	1970s	Strat. Reconn
Tu-141 (VR-2 Strizh)	Russia	Jet	11,800#	1970s	Tact. Reconn
Tu-243 (VR-3 Reyes-D)	Russia	Jet	3,080#	1980s	Tact. Reconn
Tu-300	Russia	Jet	——	2000s	Tact. Reconn
UAOS	S Africa	Piston	275#	1990s	Surv/Acquisition

UAX-X1	Turkey	Piston	540#	1990s	Surv/Acquisition
URAV	EADS	Jet	900kg	2000s	Reconn
Ugglan	France	—	—	1990s	Multi
V-1	Germany	Pulse-jet	—	1930s	Offensive
VPA-BL	France	Rocket	—	1990s	Decoy
VR-2	Russia	Jet	11,800#	—	Reconnaissance
Vampire	Canada	Jet	—	1990s	Target
Vigilant	France	Piston	88#	1980s	Multi
Vindicator	Canada	Piston	180#	1980s	Target
Vulture S	Africa	—	220#	2000s	Multi
Watchkeeper	UK	Jet	—	2000s	Multi
Wektor	Poland	Piston	—	1990s	Reconn
Yak-060	Russia	Piston	225#	1980s	Reconn/EW
Yak-061	Russia	Piston	285#	1990s	Reconn